Second Edition

Quantitative Drug Design

A CRITICAL INTRODUCTION

Second Edition

Quantitative Drug Design

A CRITICAL INTRODUCTION

YVONNE CONNOLLY MARTIN

CRC Press
Taylor & Francis Group
Boca Raton London New York

CRC Press is an imprint of the
Taylor & Francis Group, an **informa** business

CRC Press
Taylor & Francis Group
6000 Broken Sound Parkway NW, Suite 300
Boca Raton, FL 33487-2742

First issued in paperback 2019

© 2010 by Taylor and Francis Group, LLC
CRC Press is an imprint of Taylor & Francis Group, an Informa business

No claim to original U.S. Government works

ISBN-13: 978-1-4200-7099-6 (hbk)
ISBN-13: 978-0-367-38414-2 (pbk)

Library of Congress Cataloging-in-Publication Data

Martin, Yvonne Connolly, 1936-
    Quantitative drug design : a critical introduction / Yvonne C. Martin. -- 2nd ed.
        p. ; cm.
    Includes bibliographical references and index.
    ISBN 978-1-4200-7099-6 (hardcover : alk. paper)
    1. Drugs--Structure-activity relationships. 2. QSAR (Biochemistry) 3. Drugs--Design.
I. Title.
        [DNLM: 1. Drug Design. 2. Biopharmaceutics. 3. Models, Theoretical. 4.
Quantitative Structure-Activity Relationship. QV 744 M383q 2010]

    RM301.42.M37 2010
    615'.19--dc22   2009028151

Visit the Taylor & Francis Web site at
http://www.taylorandfrancis.com

and the CRC Press Web site at
http://www.crcpress.com

Contents

Chapter 4
Calculating Physical Properties of Molecules 49

Chapter 6
Form of Equations that Relate Potency and Physical Properties.................... 107

Contents   xi

Preface

Why am I revising *Quantitative Drug Design* now, thirty-some years after the first edition was written?

- First, the field has changed dramatically with the possibility to now explicitly consider 3D features in quantitative structure–activity studies. The 3D structures of the ligands might be derived from the structure of a ligand–macromolecule complex or from the new methods of pharmacophore discovery.
- Second, improvements in computer hardware and software make the methods accessible to more scientists. This ease of use spawns the need to explain the principles behind the methods and to transmit the knowledge of more experienced workers.
- Third, many scientists now use 3D searching based on the 3D structure of the macromolecular target to identify potentially active molecules. Because hundreds or thousands of molecules may be so identified, it is important to have a fast and accurate method to predict their potency and side-effect profile.

It is therefore the goal of this book to teach scientists how to apply quantitative structure–activity relationship techniques at a state-of-the-art level. My enthusiasm for writing this book arose from my experiences as referee of many manuscripts as well as from the unreliability of predictions that I have made. Examination of the faults made in logic, statistics, or interpretation of quantitative structure–activity relationships by myself or others has made me aware of some of the traps into which one can fall. It is my intent to not only introduce the methods, but also to discuss the standards of reliable work in such areas. I fully realize that in setting such guidelines, my own fallibilities become obvious. I hope that readers will inform me of their criticisms. My hope is to provide a guide for new workers so that they learn the strengths and weaknesses of each approach. Maybe they will also be inspired to improve the methods.

The arrangement of this edition is similar to the first. A new chapter has been added to introduce methods to identify the 3D structures for 3D QSAR. The chapter on statistics has been extended to include partial least squares, and the chapter on other methods has been totally rewritten to include discussion and examples of the use of multidimensional scaling, clustering, support vector machines, KNN potency prediction, and recursive partitioning. The original case studies have been expanded to include the results of reanalysis of the data with newer methods. Two additional sections have been added to the case studies: discovery of novel dopaminergics with pharmacophore mapping and CoMFA, and application of CoMFA to series in which the 3D structure of the ligand–protein complex is known. The discussion of Wisswesser Line Notation has been replaced by a description of SMILES. The extensive table of substituent constants has been deleted because of the availability of these values elsewhere.

Yvonne Connolly Martin
Waukegan, Illinois
January 30, 2009

Acknowledgments

There are many people who have helped me in my study of structure–activity relationships. I am especially grateful for the help that Corwin Hansch, Julius Taylor, and Ronald Wiegand provided in my early work; for the long-time collaboration with Ki Kim; for the help that C. Thomas Lin provided as Ki and I learned PLS and explored 3D QSAR methods; for the unfailing computer support of Jerry DeLazzer; and for the enthusiasm for the methods exhibited by my collaborators—especially Hal Jones, Tom Perun, and Bob Schoenleber. In preparing this revised edition, Stevan Djuric, Steve Muchmore, and Phil Hajduk have provided continual support and encouragement.

My special thanks go to my daughter Margaret Martin who examined the draft of the book in great detail for both clarity of presentation and the finer points of grammar and style. Her comments led to many improvements in the readability of the chapters.

My most staunch supporter is my husband; it is he who convinced me to write the original version this book and supported that effort with labor of his own at home. He and my other daughter, Gita Catherine Brown, provided encouragement when the task seemed too arduous to be worth continuing.

Yvonne Connolly Martin
Waukegan, Illinois
January 30, 2009

1 Overview of Quantitative Drug Design

As each decade passes, the techniques available for the rational design of biologically active molecules become more varied and more reliable. However, even today, we still rely on making and testing many analogs before we identify the compound whose properties allow it to progress to clinical testing in humans. This book will emphasize those computational techniques that analyze the resulting relationships between molecular structure and biological activity. Major emphasis will be given to those methods that provide a quantitative forecast of biological potency.

The practice of drug design draws upon aspects of several disciplines; physical and organic chemistry, structural biology, statistics, pharmacology, pharmacokinetics and drug metabolism, and biochemistry are perhaps the most important. The success of any one individual or a team in this task depends on how well they integrate these diverse disciplines. For example, predictions based on the careful experimental quantification of the physical properties of a set of molecules will be incorrect if the proper statistical evaluation of the relationship between these properties and potency has not been made. Alternatively, brilliant calculations on faulty biological data or incorrect chemical structures are likewise faulty. Thus, serious workers will always be alert for errors in experimental data or for gaps in their understanding of the disciplines behind drug design.

It is important to recognize that biological testing is performed on a physical sample, a compound. Usually, this compound is associated with a chemical structure, which may or may not have been correctly assigned or recorded. Additionally, the exact molecular species that triggers the biological response is usually only one of several equilibrating forms, such as different ionic states or tautomers, present in the solution. Hence, the exact molecular structures used in a computational model of structure–activity relationships (SAR) determine the model's relevance. The specific issue of tautomerization and ionization will be discussed in Chapter 6.

I. STAGES OF DRUG DISCOVERY

There are several stages in the design of a synthetic compound to be used as a drug.[1,2] Each step uses a different type and quality of information. The various computational approaches to drug design are often more applicable to one or another of these stages.

A. DISCOVERY OF THE INITIAL LEAD

After the definition of the type of biological properties that the sought-after compound should possess, the first step in the discovery of a drug of novel structure is to find a lead compound. This is usually a molecule with demonstrable but weak activity of the desired type whose structure can be modified to improve potency or other properties.

An important source of lead compounds is high-throughput screening (HTS) of either traditional compound collections or combinatorial libraries.[3,4] HTS involves the automated testing of many thousands of compounds in a simple biochemical assay. A more robust but lower-throughput test confirms or rejects the activity of compounds that appear to be active in the screen. Because the number of possible drug molecules is so large—estimated to be $>10^{180}$—computational methods are often used to select diverse compounds for screening and to design combinatorial libraries.[5] SAR by nuclear magnetic resonance (NMR) is a screening method that identifies small fragments that bind near each other in a macromolecular active site and can be linked to produce molecules with high affinity.[6]

Alternatively, HTS may be performed with virtual screening that is performed in the computer.[7,8] These methods identify those compounds from the screening library that have the best chance of being active. One approach to virtual screening uses computational methods to identify molecules that might bind to a protein of known three-dimensional (3D) structure, usually one determined by x-ray crystallography.[9] Another approach uses existing structure–activity data to form a pharmacophore hypothesis, a hypothesis as to the 3D features of a molecule that are required for biological activity, and searches a 3D database for molecules that match the hypothesis.[10–13] Either the features of the experimental binding site or a pharmacophore hypothesis can be used for manual or computer *de novo* assembly of new compounds.[14,15] Other virtual screening methods search a database for molecules that are similar in some respects to known actives.[16]

Compounds with known biological activity are often good leads. The oral antidiabetic agents, of which tolbutamide (Structure 1.1) is an example, were invented by the modification of an antibacterial sulfa drug (Structure 1.2) that lowered blood glucose as a side effect.[17] The natural anticoagulant, bishydroxycoumarin (Structure 1.3), was the lead for the synthesis of a distant analog, the widely used anticoagulant drug warfarin (Structure 1.4).[18] Natural body substances can also serve as leads. Pronethalol (Structure 1.5) was designed to antagonize the blood pressure-elevating effects of norepinephrine (Structure 1.6).[19] Also, enzyme inhibitors are often designed from the structure of substrates. For example, the monoamine oxidase inhibitor pargyline (Structure 1.7) is an analog of the substrate benzylamine (Structure 1.8).[20]

B. DEVELOPMENT OF STRUCTURE–ACTIVITY RELATIONSHIPS

The second stage in drug design involves the synthesis and testing of an exploratory series of compounds related to the lead compound. By proposing an optimal series of molecules to examine the possible physical or structural correlates of biological

(1.1)

(1.2)

(1.3)

(1.4)

(1.5)

(1.6)

(1.7)

(1.8)

potency or to probe accessory binding sites near the recognition, computational methods can decrease the number of compounds required.

C. REFINEMENT OF THE STRUCTURE–ACTIVITY HYPOTHESIS

In the third stage of drug design, one has enough structure–activity information to formulate a hypothesis of the requirements for biological activity. In the context of the emphasis of this book, this hypothesis would be a preliminary quantitative structure–activity relationship (QSAR) of the relationship between molecular properties and biological potency. It would also involve designing additional analogs to test the QSAR. At this point, one may continue to modify the structure to improve pharmacokinetic properties or to eliminate unwanted side effects, perhaps using quantitative models to forecast these properties as well.[21]

Ultimately, the potency of proposed compounds can be predicted with confidence, and the optimum of all relevant properties defined. This generates a feeling of confidence that the best analog has been synthesized. Although this stage is more or less hypothetical, the objective is to not discard a series until it has been reached.

II. COMPUTER DESCRIPTIONS OF CHANGES IN STRUCTURE RELATED TO CHANGES IN PROPERTIES OF MOLECULES

Because the analysis of structure–activity relationships is central to drug discovery, the aim of computational methods is to understand how changes in molecular structure lead to changes in molecular properties, some of which are related to changes in biological potency. There are many different approaches to describe how a change in molecular structure changes the properties of a molecule. In this section, the salient features of these molecular descriptions will be discussed to provide a framework for the discussions in subsequent chapters. Readers may also wish to consult other sources as well.[22–24]

A. SUBSTRUCTURE-BASED QSARS

Substructure-based QSAR methods are based on the hypothesis that each substructure within a molecule contributes independently to its biological potency and that the same substructure in different molecules contributes the same amount to that potency. For example, one could hypothesize that the $-SO_2NH-$ group in Structures 1.1 and 1.2 contributes equally to their potency in lowering blood glucose, or that the $-C(OH)CH_2N-$ in Structures 1.5 and 1.6 contributes equally to their affinity for the receptor for norepinephrine (Structure 1.6).

For substructure-based QSAR, each molecule is described by the presence (1.0) or absence (0.0) of a particular substructure or by the count of the number of times a substructure occurs in the molecule. For small data sets, one can identify the substructures by visual inspection. For example, in the Free–Wilson method, one assumes that, within a series of related molecules that are modified at more than one position, the effect of a particular substituent at a specific position is independent of the effect of that or any other substituent at the other positions.[25–27] This approach requires that each substituent occurs more than once at each position in which it is found.

For example, Table 1.1 shows the structure–activity relationships of erythromycin A (Structure 1.9) and B (Structure 1.10) analogs that have been studied for antibacterial potency against *Staphylococcus aureus*. The multiple regression equation that describes this data is

Erythromycin A, R=OH (**1.9**)
Erythromycin B, R=H (**1.10**)

TABLE 1.1
Structure–Activity Relationships of Erythromycin Esters

Compound	2″	4′	11	A or B	Log Potency $(\log(1/C_{SA}))$
1.09	-OH	-OH	-OH	A	3.00
1.10	-OH	-OH	-OH	B	2.82
1.11	-OC(=O)H	-OH	-OH	A	2.91
1.12	-OC(=O)H	-OH	-OH	B	2.78
1.13	-OC(=O)CH$_3$	-OH	-OH	A	2.75
1.14	-OH	-OC(=O)H	-OH	A	2.72
1.15	-OC(=O)H	-OC(=O)H	-OH	A	2.71
1.16	-OH	-OH	-OC(=O)H	B	2.71
1.17	-OC(=O)CH$_3$	-OC(=O)H	-OH	A	2.70
1.18	-OC(=O)CH$_3$	-OH	-OH	B	2.54
1.19	-OH	-OC(=O)H	-OH	B	2.54
1.20	-OC(=O)H	-OC(=O)H	-OH	B	2.50
1.21	-OH	-OC(=O)H	-OC(=O)H	B	2.42
1.22	-OC(=O)H	-OC(=O)H	-OC(=O)H	B	2.39
1.23	-OH	-OH	-OC(=O)CH$_3$	B	2.29
1.24	-OC(=O)CH$_3$	-OC(=O)H	-OH	B	2.28
1.25	-OH	-OC(=O)CH$_3$	-OH	A	2.24
1.26	-OC(=O)CH$_3$	-OC(=O)CH$_3$	-OH	A	2.15
1.27	-OC(=O)CH$_3$	-OH	-OC(=O)CH$_3$	B	2.11
1.28	-OC(=O)CH$_3$	-OC(=O)CH$_3$	-OH	B	2.04
1.29	-OH	-OC(=O)CH$_3$	-OH	B	2.02
1.30	-OH	-OC(=O)CH$_2$CH$_3$	-OH	B	2.02
1.31	-OC(=O)CH$_3$	-OC(=O)CH$_2$CH$_3$	-OH	B	1.87
1.32	-OC(=O)CH$_3$	-OC(=O)H	-OC(=O)CH$_3$	B	1.65
1.33	-OH	-OC(=O)CH$_3$	-OC(=O)CH$_3$	B	1.51
1.34	-OC(=O)CH$_3$	-OC(=O)CH$_3$	-OC(=O)CH$_3$	B	1.39
1.35	-OH	-OC(=O)CH$_2$CH$_3$	-OC(=O)CH$_2$CH$_3$	B	1.38
1.36	-OC(=O)CH$_3$	-OC(=O)CH$_2$CH$_3$	-OC(=O)CH$_2$CH$_3$	B	1.18

Source: Reprinted from Martin, Y. C.; Jones, P. H.; Perun, T.; Grundy, W.; Bell, S.; Bower, R.; Shipkowitz, N. *J. Med. Chem.* **1972**, *15*, 635–8.

$$\log(1/C_{SA}) = 2.78 + 0.21A - 0.03FO2 - 0.17AC2 - 0.27FO4 - 0.69AC4$$

$$- 0.76PR4 - 0.09FO11 - 0.56AC11 - 0.66PR11 \qquad (1.1)$$

$$R^2 = 0.986, \; s = 0.072, \; n = 28$$

In this model, C_{SA} is the relative minimum inhibitory concentration normalized to erythromycin A as 1000, A indicates that the compound is an erythromycin A (not an erythromycin B) derivative, FO indicates a formyl ester, AC an acetyl ester, and PR a propionyl ester. The numbers indicate the position of substitution. Note that the coefficients for FO2 and FO11 are not significantly different from zero and so would be omitted from the final model. R^2 indicates the fraction of variance in the data that is explained by the preceding equation—the higher the R^2, the better the fit. The value s is the standard deviation of the observed from the fitted values—the lower the s, the better the fit. This analysis will be discussed in more detail in Chapter 10, and the statistics in Chapter 7.

On the other hand, if one's data set contains diverse structures or hundreds of molecules, one would typically use an automated substructure perception program. Such programs could generate descriptors for the presence or absence of hundreds or thousands of such descriptors.[28–30] In such cases, various statistical or machine-learning methods, to be discussed in detail in Chapter 11, are used to uncover the relationship between substructures and biological activity. The result is a function that describes the contribution of each substructure to potency. The function not only summarizes the known structure–activity relationships but can also be used to fore-cast the potency of additional molecules. Automated methods for the identification of substructures will be discussed in detail in Chapter 4. The statistical and other methods are discussed in Chapters 7 and 10.

A clear advantage of substructure descriptors is that it is possible to calculate the properties of any molecule, no matter how unusual its structure may be. Hence, very diverse structures can be included within one analysis. Because the fragments are usually calculated automatically, the development of a QSAR model can be accom-plished quickly.

A disadvantage of models based on substructure descriptors is that the only information provided is the contribution of each investigated substructure to bio-activity. Because they provide no information as to the importance of any physical property to potency, it is not possible to generalize beyond the types of molecules in the model.

B. TRADITIONAL PHYSICAL-PROPERTY-BASED QSARS

Traditional QSAR assumes that the differences in biological potency within a series of molecules can be explained by differences in one or more of the physical prop-erties of these molecules. Classic studies showed the importance of differences in hydrophobicity measured as solvent–water partition coefficients[31,32] and differences in electronic properties measured as changes in pK_a or the Hammett σ constant[23,33] and redox potential.[34,35] Hence, a traditional QSAR analysis examines the quan-titative relationship between potency changes and the effect of substituents on the hydrophobicity and electronic properties of key parts of the molecule.[23,36] It also considers how the potency is affected by changes in the shape of the molecule. This results in the so-called Hansch equation[23] shown in Equation 1.2:

$$\log 1/C = a + b \log P - c (\log P)^2 + \rho \sigma + d E_s \qquad (1.2)$$

In this equation, $\log 1/C$ is the logarithm of the concentration at which the standard biological response is observed. Examples would be a pK_i for enzyme inhibition or for displacement of a standard ligand from a receptor. The coefficients a, b, c, ρ, and

d are constants that are fit by statistical analysis, some of which may be not significantly different from zero.

For reasons that will be discussed in detail in Chapter 4, log P is typically the logarithm of the octanol–water partition coefficient P defined by Equation 1.3:

$$P = \frac{[D]_n}{[D]_a} \tag{1.3}$$

The square brackets indicate concentration, the subscript a indicates the aqueous phase, and the subscript n the nonaqueous phase. The parabolic relationship with log P allows for an optimum of this property. Other forms of the relationship between potency and log P will be discussed in Chapter 6. In place of log P, one may instead use the substituent constant π_x, which describes the effect of substituent X on log P:[37]

$$\pi_X = \log P_{YX} - \log P_{YH} \tag{1.4}$$

In Equation 1.4, log P refers to the logarithm of the octanol–water partition coefficient and Y is any appropriate parent structure.

The σ is the classic Hammett constant for the electronic effect of a substituent.[33] It is defined as the effect of the substituent on the pK_a of benzoic acid (Structure 1.37).

$$\sigma_X = \log K_{a,X} - \log K_{a,H} \tag{1.5}$$

Chapter 4 will discuss calculating the electronic effects of substituents in more detail.

(1.37)

Similar to the Hammett constant, E_s models the steric effect of a substituent.[23,38] It is measured as the effect of an acyl substituent on the rate of hydrolysis of the corresponding ethyl ester (Structure 1.38).

$$E_{s,X} = \log k_{XCH_2CO_2R} - \log k_{CH_3CO_2R} \tag{1.6}$$

(1.38)

$k_{XCH_2CO_2R}$ is the rate constant for the hydrolysis of the substituted ester, and $k_{CH_3CO_2R}$ is that for the hydrolysis of the acetate ester.

The structures of the molecules to be analyzed may require somewhat different descriptors for the effects described in Equation 1.2. Further definition and estimation of descriptors for traditional QSAR will be discussed in detail in Chapter 4.

As an example of the application of Equation 1.2, Model 1.7 provides an equation that relates the antibacterial activity of the erythromycin esters with log P:

$$\log(1/C) = 6.21 - 1.25 \log P \tag{1.7}$$

$$R^2 = 0.79, \quad s = 0.237, \quad n = 28$$

The higher R^2 and lower s of Model 1.8 show that a much closer fit is obtained by adding substructure descriptors D4 and D11 to indicate esterification at the 4″ and 11 positions, respectively, and A to indicate that the compound is an erythromycin A derivative:

$$\log(1/C) = 6.89 - 1.36 \log P - 0.29\,D4 - 0.17\,D11 - 0.36A \tag{1.8}$$

$$R^2 = 0.95, \quad s = 0.127, \quad n = 28$$

Chapter 10 will discuss the insights gained by comparing Models 1.1, 1.7, and 1.8 and other erythromycin data sets.

A QSAR model provides a mental picture of the pertinent intermolecular interactions of the ligands in the biological system. We may not understand a hydrophobic interaction, but if our biological response is positively correlated with log P and not pK_a, we form a mental picture that hydrophobic biomolecules interact with the various ligand molecules, and that whatever acid–base reactions occur influence all analogs equally.

Traditional QSAR is relatively easy and inexpensive to use. The physical properties, especially log P, are often readily calculated, and the statistics can be determined in programs such as Microsoft Excel.

An important advantage of QSAR is that hydrophobic effects of molecules are often the driving force for the strength of their interaction and that it is straightforward to calculate or measure log P. As a result, many QSAR equations have been published[23] and thousands are available in a database.[39]

The final advantage of QSAR is that the conclusions reached have application beyond the substituents included in the analysis, and perhaps beyond the particular structure class calculated. For example, one might observe similar optimum log P's for a series of varying structures but similar mode or site of action.[40]

A problem with traditional QSAR is that one must be able to calculate the properties of interest for the substituents in the data set. Although some missing values can be estimated, often compounds must be omitted from the analysis simply because one cannot estimate the pertinent properties.

The most serious limitation of QSAR is that the experimental models that provide the basis for the descriptors for QSAR are imperfect models for interactions in the biological system. In particular, steric interactions are extremely difficult to extrapolate from system to system. One usually is not sure of the atom from which to estimate steric effects. Second in obscurity are electronic effects. Again, part of the problem is to decide the key atom from which substituent effects should be estimated.

C. QSARs BASED ON 3D PROPERTIES OF MOLECULES

Because the target biomolecule and the various ligands exist in three dimensions, in the search for quantitative structure–activity relationships, it is logical to consider the 3D molecular properties of the ligands. This is accomplished in comparative molecular field analysis (CoMFA)[41] and newer methods. Unlike traditional QSAR, 3D QSARs does not use previously measured values of substituent constants, but instead, uses molecular descriptors calculated from the 3D structures of the molecules themselves—the electrostatic and steric fields surrounding each molecule in the series, for example, calculated by sampling the region around each molecule at the intersections of a 1 or 2 Å lattice.

CoMFA and most newer 3D QSAR methods require that one superimpose the molecules before the calculation. Experimental or in silico studies of the structure of macromolecule–ligand complexes can provide this information.[42] Alternatively, various computer programs are designed to help one choose the proposed bioactive conformation and the superposition rule.[12] Molecular modeling as a prelude to 3D QSAR will be discussed in more detail in Chapter 3.

Because there are typically thousands of sampling points around the molecules, one cannot use traditional regression analysis to uncover the relationships between biological potency and the values of the fields at the various points. Instead, the statistical technique of partial least squares (PLS) is used. The result is a function that can be contoured for a 3D display and also be used to forecast the affinity of additional compounds. Figure 1.1 shows a 3D contour plot of the CoMFA analysis of the erythromycin esters discussed earlier. Chapter 7 will discuss PLS in more detail; Chapters 9 and 10 will discuss the analysis of the structure–activity relationships of erythromycin analogs.

Chapter 10 will provide several examples of using the 3D structures of the protein–ligand complexes to align the molecules for 3D QSAR. We found that CoMFA based on these alignments fits the structure–activity relationships much more precisely than do more sophisticated energy functions and complex calculations. This result is important in the light of the challenges in finding a general function that accurately predicts ligand–protein binding affinity.[9]

Although the methodologies are different, Kim has shown that traditional QSAR and CoMFA describe the same properties of the molecules.[43–45] In some cases, traditional and 3D descriptors are included in one analysis.[46]

3D QSAR continues to be an area of intensive research. Since the original description of CoMFA, hundreds of applications, improvements to the methodology, and several other approaches to 3D QSAR have been reported.[47–50] For example, comparative molecular similarity indices analysis (CoMSIA) places the grid inside the union surface of the molecules,[51] CoMBINE calculates interactions of the bound ligand with the target protein,[52] and 4D QSAR uses statistics to identify the correct superposition rule as well as to forecast potency.[53]

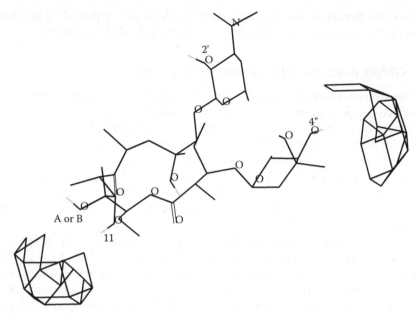

FIGURE 1.1 The negative steric contours generated from a CoMFA analysis of the data in Table 1.1. The model has an R^2 of 0.939 and a standard error of fit of 0.129 (see Chapter 10 for further discussion).

Catalyst[54] and Phase[55] are 3D QSAR methods that do not use a lattice of energy values, but instead locate small regions in space, the occupancy of which with appropriate atoms or projections from atoms is quantitatively related to potency. These methods also identify the conformation to be used for each molecule and establish the superposition rule by statistical analysis.

A great strength of 3D QSARs is that they provide a visual display that shows favorable and unfavorable regions for specific molecular properties. These displays provide a direct visual summary of the structure–activity relationships within the set of molecules.

A second advantage of this family of QSARs is that any molecule can be included because the properties are calculated using formulas that are well established in molecular modeling. Hence, one can use 3D QSAR for any set of molecules for which one can propose a superposition rule or has one that was established experimentally.

A frustration with 3D QSAR methods that require molecular superposition is that it may be difficult or very time consuming to decide on a conformation and superposition rule. This issue will be discussed further in Chapter 3. Additionally, CoMFA results may be affected by the position of the molecules relative to the lattice, the lattice spacing, and the lattice.[56]

3D QSARs suffer from the problem that the statistics must be carefully performed, with the result that extensive calculations must be done to show that the results are not chance correlations.

III. LESSONS LEARNED

This overview highlights the strategies by which computational methods produce quantitative predictions of the potency of novel molecules. The exact strategy to arrive at such a model will depend on the information available to the modeler—if one has structures of the target with at least some of the ligands bound or if the ligands are conformationally constrained, a 3D QSAR method is probably the method of choice. On the other hand, 2D methods may be more appropriate if it appears that physical properties of the molecules are responsible for varying potency, or if there is no 3D information to guide the selection of the bioactive conformation or how to superimpose the molecules. Because molecular structure determines molecular properties, 2D and 3D molecular descriptors are not independent but, rather, they are alternative descriptors of the same fundamental properties.

REFERENCES

1. Ramsden, C. A., Ed. *Comprehensive Medicinal Chemistry: The Rational Design, Mechanistic Study, and Therapeutic Application of Chemical Compounds*. Pergamon: Oxford, 1990.
2. Wermuth, C. G., Ed. *The Practice of Medicinal Chemistry*. Academic Press: London, 1996.
3. Green, D.; V. S. *Prog. Med. Chem.* **2003**, *41*, 61–97.
4. Irwin, J. J.; Shoichet, B. K. *J. Chem. Inf. Model.* **2005**, *45*, 177–82.
5. Willett, P. *Curr. Opin. Biotechnol.* **2000**, *11*, 85–8.
6. Shuker, S. B.; Hajduk, P. J.; Meadows, R. P.; Fesik, S. W. *Science* **1996**, *274*, 1531–4.
7. Klebe, G. *Drug Discovery Today* **2006**, *11*, 580–94.
8. McGaughey, G. B.; Sheridan, R. P.; Bayly, C. I.; Culberson, J. C.; Kreatsoulas, C.; Lindsley, S.; Maiorov, V.; Truchon, J. F.; Cornell, W. D. *J. Chem. Inf. Model.* **2007**, *47*, 1504–19.
9. Warren, G. L.; Andrews, C. W.; Capelli, A. M.; Clarke, B.; La Londe, J.; Lambert, M. H.; Lindvall, M.; Nevins, N.; Semus, S. F.; Senger, S.; Tedesco, G.; Wall, I. D.; Woolven, J. M.; Peishoff, C. E.; Head, M. S. *J. Med. Chem.* **2006**, *49*, 5912–31.
10. Martin, Y. C. *J. Med. Chem.* **1992**, *35*, 2145–54.
11. Güner, O. F., Ed. *Pharmacophore Perception, Development, and Use in Drug Design*. International University Line: La Jolla, CA, 1999.
12. Martin, Y. C. In *Comprehensive Medicinal Chemistry* II; Mason, J. S., Ed. Elsevier: Oxford, 2007, Vol. 4, pp. 119–47.
13. Martin, Y. C. In *Comprehensive Medicinal Chemistry* II; Mason, J. S., Ed. Elsevier: Oxford, 2007, Vol. 4, pp. 515–36.
14. Bohm, H. J. *Perspect. Drug Discovery Des.* **1995**, *3*, 21–33.
15. Gillet, V. J.; Willett, P.; Bradshaw, J. *J. Chem. Inf. Comput. Sci.* **1997**, *37*, 731–40.
16. Schneider, G.; Neidhart, W.; Giller, T.; Schmid, G. *Angew. Chem.*, Int. Ed., **1999**, *38*, 2894–6.
17. Grunwald, F. A. In *Med. Chem.*, 3rd ed.; Burger, A., Ed. Wiley: New York, 1970, pp. 1172–84.
18. Cutting, W. *Handbook of Pharmacology*, 4th ed. Appleton-Century-Crofts: New York, 1969.
19. Barrett, A. M. In *Drug Design*, 4th ed.; Ariens, E. J., Ed. Academic Press: New York, 1972, Vol. 3, pp. 205–28.
20. Taylor, J. D.; Wykes, A. A.; Gladish, Y. C.; Martin, W. B. *Nature* **1960**, *187*, 941–2.

21. Ekins, S. *Biochem. Soc. Trans.* **2003**, *31*, 611–14.
22. Kubinyi, H. *QSAR: Hansch Analysis and Related Approaches.* VCH: Weinheim, Germany, 1993, Vol. 1.
23. Hansch, C.; Leo, A. *Exploring QSAR: Fundamentals and Applications in Chemistry and Biology.* American Chemical Society: Washington, DC, 1995.
24. Perkins, R.; Fang, H.; Tong, W. D.; Welsh, W. *J. Environ. Toxicol. Chem.* **2003**, *22*, 1666–79.
25. Free, S. M.; Wilson, J. *J. Med. Chem.* **1964**, *7*, 395–9.
26. Craig, P. N. In *Biological Correlations—The Hansch Approach*; Gould, R. F., Ed. American Chemical Society: Washington, DC, 1972, pp. 115–29.
27. Kubinyi, H. In *Comprehensive Medicinal Chemistry;* Hansch, C., Sammes, P. G., Taylor, J. B., Eds. Pergamon: Oxford, 1990, Vol. 4, pp. 589–643.
28. Young, S. S.; Hawkins, D. M. *SAR QSAR Environ. Res.* **1998**, *8*, 183–93.
29. Roberts, G.; Myatt, G. J.; Johnson, W. P.; Cross, K. P.; Blower, P. E. *J. Chem. Inf. Comput. Sci.* **2000**, *40*, 1302–14.
30. Xia, X. Y.; Maliski, E. G.; Gallant, P.; Rogers, D. *J. Med. Chem.* **2004**, *47*, 4463–70.
31. Overton, E. *Z. Phys. Chem.* **1897**, *22*, 189–209.
32. Meyer, H. *Archiv fur Experimentell Pathologie und Pharmakologie* **1899**, *42*, 109–18.
33. Hammett, L. *Physical Organic Chemistry.* McGraw-Hill: New York, 1970.
34. Fieser, L. F.; Berliner, E.; Bondhus, F. J.; Chang, F. C.; Dauben, W. G.; Ettlinger, M. G.; Fawaz, G.; Fields, M.; Fieser, M.; Heidelberger, C.; Heymann, H.; Seligman, A. M.; Vaughan, W. R.; Wilson, A. G.; Wilson, E.; Wu, M.-i.; Leffler, M. T.; Hamlin, K. E.; Hathaway, R. J.; Matson, E. J.; Moore, E. E.; Moore, M. B.; Rapala, R. T.; Zaugg, H. E. *J. Am. Chem. Soc.* **1948**, *70*, 3151–5.
35. Albert, A. *Selective Toxicity*, 5th ed. Chapman and Hall: London, 1973.
36. Hansch, C.; Leo, A.; Hoekman, D. *Exploring QSAR: Hydrophobic, Electronic, and Steric Constants.* American Chemical Society: Washington, DC, 1995.
37. Fujita, T.; Iwasa, J.; Hansch, C. *J. Am. Chem. Soc.* **1964**, *86*, 5175–80.
38. Taft, R. W. In *Steric Effects in Organic Chemistry*; Newman, M. S., Ed. Wiley: New York, 1956, pp. 556–675.
39. Leo, A.; Hansch, C.; Hoekman, D. Bio-Loom, *BioByte* **2008**.
40. Kurup, A.; Garg, R.; Hansch, C. *Chem. Rev.* **2001**, *101*, 2573–600.
41. Cramer III, R. D.; Patterson, D. E.; Bunce, J. D. *J. Am. Chem. Soc.* **1988**, *110*, 5959–67.
42. Machius, M. *Curr. Opin. Nephro Hypertens.* **2003**, 12, 431–8.
43. Kim, K. H. *Med. Chem. Res.* **1991**, *1*, 259–64.
44. Kim, K. H.; Martin, Y. C. *J. Med. Chem.* **1991**, *34*, 2056–60.
45. Kim, K. H. In *Trends in QSAR and Molecular Modelling '92*; Vermuth, C. G., Ed. ESCOM: Leiden, 1993, pp. 245–51.
46. Greco, G.; Novellino, E.; Silipo, C.; Vittoria, A. *Quant. Struct–Act. Relat.* **1992**, *11*, 461–77.
47. Kubinyi, H., Ed. *3D QSAR in Drug Design: Theory Methods and Applications.* ESCOM: Leiden, The Netherlands, 1993.
48. Greco, G.; Novellino, E.; Martin, Y. C. In *Reviews in Computational Chemistry*; Lipkowitz, K. B., Boyd, D. B., Eds. VCH: New York, 1997, Vol. 11, pp. 183–240.
49 Kubinyi, H.; Folkers, G.; Martin, Y. C., Ed. *3D QSAR in Drug Design*, Vol. 2, Ligand–Protein Interactions and Molecular Similarity. ESCOM: Leiden, The Netherlands, 1998.
50. Kubinyi, H.; Folkers, G.; Martin, Y. C., Ed. *3D QSAR in Drug Design*, Vol. 3 *Recent Advances.* ESCOM: Leiden, The Netherlands, 1998.
51. Klebe, G.; Abraham, U.; Mietzner, T. *J. Med. Chem.* **1994**, *37*, 4130–46.

52. Ortiz, A. R.; Pisabarro, M. T.; Gago, F.; Wade, R. C. *J. Med. Chem.* **1995**, *38*, 2681–91.
53. Hopfinger, A. J.; Wang, S.; Tokarski, J. S.; Jin, B. Q.; Albuquerque, M.; Madhav, P. J.; Duraiswami, C. *J. Am. Chem. Soc.* **1997**, *119*, 10509–24.
54. Sprague, P. W. *Perspect. Drug Discovery Des.* **1995**, *3*, 1–20.
55. Dixon, S. L.; Smondyrev, A. M.; Knoll, E. H.; Rao, S. N.; Shaw, D. E.; Friesner, R. A. *J. Comp-Aid. Mol. Des.* **2006**, *20*, 647–71.
56. Bucholtz, E. C.; Tropsha, A. *Med. Chem. Res.* **1999**, *9*, 675–85.

2 Noncovalent Interactions in Biological Systems

Usually the binding of a ligand to its target biomolecule does not involve covalent bond formation; rather, it is the result of several noncovalent interactions.[1–3] Furthermore, even if a covalent bond is formed, at least part of the recognition and specificity between a ligand and its target is governed by noncovalent interactions. Hence, to understand the forces that influence the strength of a ligand–target interaction, we need to understand the types and energetics of noncovalent interactions, which is the subject of this chapter. The relative strengths of the various types of interactions, the contributions that they can make to specificity, and how changes in molecular structure change the strength of these interactions will be covered. This will provide a foundation for wisely choosing and calculating descriptors for a quantitative structure–activity relationship (QSAR), which will be discussed in Chapter 4.

Six distinct types of noncovalent interactions may be present in a ligand–macromolecule complex:

1. Hydrogen bonds are characterized by the close approach between a polar atom and a hydrogen atom that is covalently bonded to another polar atom. Usually the three atoms lie on a straight line. For example, the hydrogen bond between a carbonyl oxygen and the hydrogen atom of a hydroxyl group can be diagrammed as $>C=O\cdots H\text{-}O\text{-}$.
2. Electrostatic forces result from the interaction of formally charged groups of opposite signs, for example, a carboxylate ($\text{-}C(=O)O^-$) and a hydrogen atom of a basic primary amine ($\text{-}NH_4^+$).
3. Hydrophobic interactions are governed by the tendency of groups such as hydrocarbon chains to interact with each other rather than with the surrounding water. The driving force is the strong tendency of water molecules to interact with each other.
4. Dispersion interactions result from the momentary shift of the electron cloud of one group with the slightly oppositely charged electron cloud of another group.
5. Charge-transfer interactions, as the name implies, result from the transfer of charge from one group, an electron donor, to another, an electron acceptor, to form a complex.

6. Steric repulsion results from the fact that two atoms cannot occupy the same region in space. Each atom has a radius that defines a surface into which no other atom can penetrate.

For example, Figure 2.1 shows a LIGPLOT diagram[4] of the interaction of inhibitor A-250061 (Structure 2.1) with urokinase.[5] There are electrostatic interactions of the carboxylates of Asp 191 and Asp 49 with basic nitrogen groups on the inhibitor; hydrogen bonds between Ser 192, Gly 220, Gln 194,

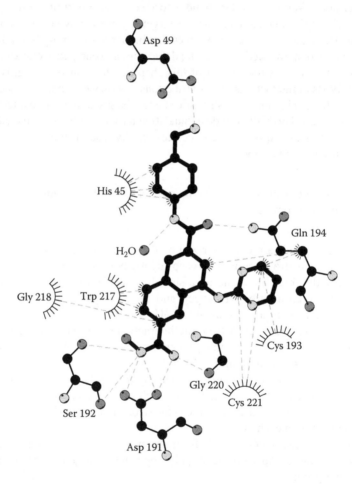

Key

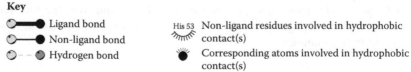

FIGURE 2.1   A LIGPLOT[4] diagram of the interactions between A-250061 and urokinase.[5]

(2.1)

and water with polar atoms of the inhibitor; and hydrophobic interactions of Cys 193, Cys 221, Trp 217, Gly 218, His 45, and Gln 194 with nonpolar atoms of the inhibitor.

I. FACTORS THAT INFLUENCE THE STRENGTH OF AN INTERACTION

How does a change in structure of a ligand change the equilibrium constant of an interaction between a ligand L and a macromolecule M?

$$L + M \xrightleftharpoons{K_{eq}} LM \qquad (2.1)$$

$$K_{eq} = \frac{[LM]}{[L] \times [M]} \qquad (2.2)$$

Thermodynamics teaches us that the change in the equilibrium constant depends on the change in the heat, the enthalpy of the system, and the enthalpy's order/disorder. This is stated formally in the fundamental thermodynamic relationship:

$$-RT \ln K_{eq} = \Delta G^0 = \Delta H^0 - T \Delta S^0 \qquad (2.3)$$

in which R is the gas constant; T, the temperature of measurement; ΔG^0, the change in Gibb's free energy accompanying the interaction; ΔH^0, the change in enthalpy; and ΔS^0, the change in entropy of reaction.[6] The superscripts denote that the thermodynamic quantities are measured at the standard state.

The ΔH^0 term includes the potential energy of the complex ΔE_p^0, which describes the energies of hydrogen bond formation, electrostatic interaction, dispersion, and charge-transfer interactions. However, this is only one component of ΔH^0:

$$\Delta H^0 = \Delta E_p^0 + \Delta E_z + \int_0^t \Delta C_p dT \qquad (2.4)$$

ΔH^0 also includes ΔE_z, the change in zero-point energy at $0°K$, and ΔC_p, the change in heat capacity.[7] The latter two terms are usually ignored in discussions because they are assumed to be constant within a series or proportional to ΔE_p^0.

(2.2)   **(2.3)**

The ΔS^0 term, the change in entropy due to molecular interactions, is related to the change in the orderliness of the total system including the solvent, isolated components, and the complex. An increase in entropy corresponds to increasing disorder in the system. From Equation 2.3, we see that this leads to an increase in the equilibrium constant or a stronger association in the complex. In terms of entropy, spontaneous reactions are those in which the system becomes more disordered.

In summary, Equations 2.3 and 2.4 show that the changes in an equilibrium or rate constant that result from changes in structure are a complex function of changes in potential energy, vibrational energy, heat capacity, and entropy effects. We shall see in the following text examples where changes in the equilibrium or rate constants are dominated by changes in enthalpy and also where changes are dominated by changes in entropy.

Because noncovalent interactions are generally weak compared to covalent bonds, several types of interactions often cooperate in the formation of a ligand–macromolecule complex.[8] To a first approximation, the enthalpy for the various sorts of interactions are additive. However, once any type of interaction has occurred, much of the translational entropy of the small molecule has been lost, and the entropy loss in formation of a second interaction will be much less than the first, leading to a cooperative effect. Such cooperative interactions mean that several interactions that are individually weak or inconsequential may add up to a strong overall interaction.

Because several different types of interactions are involved, cooperativity also leads to the selectivity characteristic of ligand–macromolecule interactions. An example of cooperativity is seen with serum albumin. It binds formate (Structure 2.2) with a ΔG^0 value of ~4 kcal/mol, and binds caproate (Structure 2.3) with a ΔG^0 value of 7 kcal/mol.[9] This increase in ΔG^0 of binding corresponds to an increase in the equilibrium constant of 273-fold. Not only is there stronger binding with caproate than with formate but the additional interaction of caproate's hydrophobic tail illustrates that the binding site is specific for hydrophobic anions.[9]

II. THE IMPORTANCE OF WATER

Although many aspects of noncovalent interactions can be described qualitatively, the special properties of water often make it difficult to quantitate the energies involved. For this reason, measured properties are often used as models for the quantitation.

Any biological system is predominantly water; hence, knowledge of the properties of water is key to understanding its influence on the intermolecular forces that are involved in the interaction between a ligand and a macromolecule. In any type of interaction between solutes L and M, the reaction to be considered is not

$$L + M \rightleftharpoons LM \qquad\qquad (2.1)$$

but rather

$$L \cdots S_i + M - S_j \rightleftharpoons LM \cdots S_k + \frac{i+j-k}{2} S \cdots S \qquad (2.5)$$

in which S represents the solvent, and the dotted line a noncovalent interaction. That is to say, the influence of the solvent must be considered in any type of interaction. In particular, water modulates most types of intermolecular interactions.

The solvent water is unique.[10–13] In ordinary ice, the water molecules are arranged in a tetrahedral array such that each oxygen atom is surrounded by four hydrogen atoms: the two to which it is covalently bonded, and two to which it is hydrogen bonded. In ice, the water molecules and their associated hydrogen bonds are relatively fixed in space.

The same, dense, tetrahedral array of water molecules persists in liquid water. The difference between water and ice is that in liquid, the water molecules vibrate (librate) with greater amplitude along the H-bond axes.[13] As a result, concerted changes in H-bonding patterns occur. Experimental measurements suggest that the structure of liquid water is not fixed in time, but that the hydrogen bond network is continually being broken and reformed with different pairings of water molecules. These changes occur within less than a nanosecond, 10^{-9} s.[13–18]

The strength of solute–solute interactions in water is dominated by the strong intermolecular forces between water molecules and by water's ability to solvate other molecules. This ability arises from water's large dipole moment, its polarizability, its small size, and its hydrogen bonding character.[13] To complicate matters, the structure of liquid water changes in response to dissolved solutes.[19–21] Water molecules tend to orient their dipoles to neutralize the charge of dissolved ions, whereas around apolar molecules they tend to form local "icebergs."[22] Because its hydrogen bonds are rapidly changing its hydrogen bonding partners, liquid water quickly changes this network to accommodate whichever new solute is introduced. The new network is that of minimum free energy. For different solutes, even in a homologous series, entropy and enthalpy may not change proportionally.

III. ELECTROSTATIC INTERACTIONS

Several types of noncovalent interactions between or within molecules can be described as interactions between charges. They can be either charges fully developed into ions or partial atomic charges (q) that result from charge separation within an overall electrically neutral molecule, producing a dipole. Three cases will be discussed: ion–ion, ion–dipole, and dipole–dipole interactions.

The physics of electrostatic interactions is relatively clear-cut. Complications arise when one attempts to quantify the potential energy of an electrostatic interaction of a molecule with a protein or nucleic acid, because the dielectric constant of the binding site or the competing interaction in water is ambiguous. Accordingly, the equations quoted in the following text are only indicative of some of the factors that influence the strength of the interaction.

The strong interaction between two oppositely charged ions is familiar to every beginning chemist; the low volatility of sodium chloride is a simple example. The potential energy of interaction between two charges, E, relative to the energy at infinite separation is given by the following equation:

$$E = \frac{e_a e_b}{Dr} \tag{2.6}$$

where e_a and e_b are the magnitudes of the charges (either fully developed or partial atomic charges that result from polarization within the molecule), r is the distance between them, and D is the dielectric constant of the material through which the charges interact. The energy is inversely proportional to this dielectric constant, which is not necessarily that of the bulk solution. The interaction energy is favorable for ions of opposite charges, but unfavorable, or repulsive, for ions of the same charge.

If the structure of a ligand is changed by adding a substituent that delocalizes a charge, the strength of interaction is decreased because with a diffuse charge, the average distance between the charges is larger. Substituents might also sterically prevent close approach of the charges and thus decrease the electrostatic interaction energy.

Although the classic salt of a cation with an anion, sodium chloride, is not volatile, it is extremely soluble in water, in which the ion pairs completely dissociate. Why? The energy of interaction between an ion and a dipole such as water is given by the following equation:

$$E = \frac{e_a \mu \cos \Theta}{Dr^2} \tag{2.7}$$

In this equation, μ is the dipole moment of the neutral substance b (water, in this example), and Θ is the angle between the line joining the charge and the middle of the dipole and that between the ends of the dipole. In this case, r is the distance between the ion and the middle of the dipole; it is assumed to be large compared to the distance between the ends of the dipole. Ionic substances are soluble in water if the ion–dipole interactions in solution are strong enough to effectively compete with the ion–ion interactions in the crystal.

From Equation 2.7, we see that the energy of ion–dipole interactions fall off with the square of the distance, whereas Equation 2.6 shows that ion–ion interactions fall off with the first power of distance. Hence, ion–dipole interactions are more sensitive to exact spatial arrangement of the interacting partners than are ion–ion interactions. Conversely, ion–ion interactions act over a longer distance.

The energy of interaction between two dipoles is given by the following equation:

$$E = \frac{2\mu_1 \mu_2 \cos(\Theta_1) \cos(\Theta_2)}{Dr^3} \tag{2.8}$$

Note that dipole–dipole interactions fall off with the third power of distance and, hence, are more specific than are interactions involving ions.

When ions or polar molecules dissolve in water, the structure of the water changes. The complementary end of the water dipole interacts with the solute ion or dipole, sometimes forming a tight hydration layer around the solute. The structuring effect of the solute on the water may occur through several layers of solvent molecules. The net effect on the whole solution may be to either increase or decrease the organization of the solution, that is, to either decrease or increase the entropy of the system compared to that of the isolated components.

If we assume that, for electrostatic interactions, the equilibrium constant K is determined principally by ΔE_p, then ion–ion interactions are not energetically important in aqueous solution because of competition with the strong ion–water interaction. In nonaqueous solutions or the hydrophobic interior of proteins, oppositely charged ions do tend to associate. In an exactly parallel manner, ion–dipole and dipole–dipole interactions between solutes are not energetically important in aqueous solutions, but may become so in nonaqueous phases or the interior of proteins.

The effect of substituents on the partial net charge of an atom is described by Hammett σ values. Partial atomic charges may also be estimated by empirical or quantum chemical calculations. Because dipoles result from the differences in charges of atoms, Hammett σ values or empirical or quantum chemical calculations are also useful for estimating the substituent effects on dipoles.

IV. HYDROGEN BONDS

In the covalent bond between a hydrogen atom and an electronegative atom, such as -F, -O, or -N, the attraction of the electrons by the electronegative atom produces a partial positive charge on the attached hydrogen atom. This partially positive hydrogen atom can, in turn, interact with another electronegative atom, either in another molecule or in the same molecule. For example,

$$\tag{2.9}$$

In this description, the hydrogen bond is electrostatic. Many computer programs model hydrogen bonding using an electrostatic term, sometimes including a decreased radius for the polar H atom to allow it to come closer to the electronegative hydrogen bond acceptor. A covalent interpretation is also possible.[23] It is most evident in extremely short and strong homonuclear (O-H···O, N-H···N, F-H···F) hydrogen bonds. Consider the following hydrogen bonds within the HFH$^-$ ion:

$$F^- \!-\! H \!-\! F \leftrightarrow F^- \!-\! H^+ \!-\! F^- \leftrightarrow F \!-\! H \!-\! F^- \tag{2.10}$$

The covalent interpretation supports the observation that there is a preferred torsion angle between the three atoms involved in the hydrogen bond and the atom bearing the hydrogen bond acceptor.[24,25]

The hydrogen bond is unique to hydrogen because it is the only atom that (1) can carry a positive charge while covalently bonded in a molecule and (2) is small enough to allow a close approach of a second electronegative atom.

Crystallographers have studied the geometry of hydrogen bonds by investigating how molecules pack in crystals of small molecules.[24–28] Small-molecule crystals are studied in preference to protein–ligand complexes, because they are known to much higher resolution, and they represent a much larger variety of functional-group interactions. Although there is a definite bias to form a hydrogen bond along the lone-pair axis of the hydrogen bond acceptor, there is considerable deviation from this line.[25,28] The closer the two electronegative atoms approach, that is, the stronger the hydrogen bond, the more likely it is that the approach of the hydrogen atom is along the lone-pair axis. This confirms the covalent nature of these strong hydrogen bonds.[23]

Because of the possible competition with hydrogen bonding to water, it is difficult to predict the strength of any particular hydrogen bond from first principles. As might be anticipated from the discussion on electrostatic interactions, in aqueous solution solute–solute hydrogen bonds are effectively masked by the competing solute–solvent hydrogen bonds.[29] In nonpolar solvents and the hydrophobic interior of globular proteins, however, hydrogen bonds can contribute appreciably to the energy of inter- or intramolecular interactions. For example, in carbon tetrachloride, the ΔG for N-methylacetamide to form an intermolecular hydrogen bond is -0.92 kcal/mol.[29]

Changes in structure change the strength of a hydrogen bond to a common donor or acceptor in direct proportion to the effect of the structural modification on the partial atomic charge, q, of the electronegative atoms that are involved in the hydrogen bond. Hammett σ values or partial atomic charges from empirical or quantum chemical calculations on the three-dimensional structure quantitate such substituent effects. Of course, attention must be paid to the possible steric interference with the ideal hydrogen bond geometry.

The relationship between σ constants and the strength of hydrogen bonds is illustrated by the strength of the hydrogen bond in carbon tetrachloride, between substituted phenols (Structure 2.4) and pyridines (Structure 2.5), as shown in Figure 2.2. Substituents that decrease the partial atomic charge (q) of the oxygen of the phenol decrease its affinity for the proton, and thereby increase the strength of the hydrogen bond. There is a positive correlation between the σ of a substituent and its effect on the log K of association.[30] In parallel, substituents that increase the q of the pyridine nitrogen increase the strength of the hydrogen bond. A subtlety in this relationship is that a better correlation is obtained if one uses σ values in which resonance contributions are attenuated. Thus, resonance effects may be more important in the determination of the pK_a value of a phenol than in determining its hydrogen-bond-donating ability. There are other works that discuss substituent effects on hydrogen bonding in more detail.[31–34]

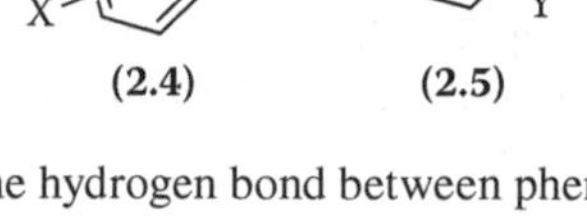

(2.4)   (2.5)

FIGURE 2.2   A diagram of the hydrogen bond between phenols and pyridines.

Several lines of evidence suggest that the hydrogen-bonding character of ligands influences the rate at which they traverse the membranes of the gastrointestinal tract[35–40] and other tissues.[41–43]

V. DISPERSION INTERACTIONS

Dispersion or London forces are those forces that, in the absence of ionic, dipolar, charge-transfer, or hydrogen bonds, hold nonpolar molecules together in the liquid phase. For example, a change in these forces is the explanation for the change in the boiling point of alkanes with changes in structure:

$$\text{Hexane}_{\text{liquid}} \rightleftharpoons \text{Hexane}_{\text{gas}} \tag{2.11}$$

Even in molecules with no permanent dipole, the vibration of electrons with respect to the nucleus results in an instantaneous dipole. The source of the dispersion energy of interaction is the tendency of such an instantaneous dipole to induce a dipole in a neighboring molecule. Hence, this type of interaction is sometimes called an *induced dipole–induced dipole interaction*. This cohesive potential energy, E, between two substances a and b is given by the following equation:

$$E = -\frac{3\alpha_a \alpha_b I_a I_b}{2r^6(I_a + I_b)} \tag{2.12}$$

In this equation, α is the polarizability of the substance, I is its ionization potential, and r is the distance between the two interacting species. Recall that the ionization potential is the lowest energy required to remove an electron from a substance.

Molar refractivity (MR) describes the possibility of molecules to participate in a dispersion interaction. Equation 2.13 shows how MR is related to α:

$$\text{MR} = \frac{4\pi N\alpha}{3} \tag{2.13}$$

where N is Avogadro's number, and π is the usual irrational real number whose value is 3.1416. Hence, the energy of interaction becomes

$$E = -\frac{K\,\text{MR}_a\,\text{MR}_b I_a I_b}{r^6(I_a + I_b)} \tag{2.14}$$

where K is a proportionality constant. Thus, the potential energy of a dispersion interaction is a complex function of the ionization potentials and molar refractivities of the two substances. Because of the r^6 dependence of dispersion interactions on distance, dispersion interactions are a significant source of specificity in ligand–macromolecule interactions. The r^6 dependence also means that identical substituents at different positions on a molecule contribute differently to E.

For a series of analogs, a, that reacts with a common target, b, at the same distance, r, the relative energy of interaction reduces to

$$E = \frac{Z \mathrm{MR}_a I_a}{I_a + I_b} \tag{2.15}$$

where Z is the proportionality constant:

$$Z = -\frac{27 \mathrm{MR}_b I_b}{32\pi^2 N^2 r^6} = \frac{K \mathrm{MR}_b I_b}{r^6} \tag{2.16}$$

If the ionization potential of the target is substantially less than that of the ligand ($I_a \gg I_b$), then Equation 2.15 reduces to

$$E = Z \mathrm{MR}_a \tag{2.17}$$

If, on the other hand, the ionization potential of the ligand is substantially less than that of the target ($I_a \ll I_b$), then

$$E = Z' \mathrm{MR}_a I_a \tag{2.18}$$

where Z' is equal to Z/I_b. Thus, often the strength of dispersion binding is proportional to either the molar refractivity or molar refractivity times the ionization potential of the variable substance.

A classic example of dispersion binding involves substances that block the electrical stimulation of excitable membranes[44]—a measure of local anesthetic effect. The logarithm of the minimum effective concentration of such substances is a linear function of the product of its ionization potential and its molar polarizability.

One may estimate the ionization potential using the energy of the highest occupied molecular orbital, HOMO, derived from quantum chemical calculations. It is often correlated with the σ constant. Within a series, ionization potential may vary much less than MR; sometimes, it is assumed to be constant. MR is an additive constitutive property that is easily calculated from available substituent constants.[45]

Since MR is an additive property, its value increases with molecular weight and, hence, overall volume. As a result, MR also reflects molecular size; negative correlations of biological potency with MR reflect steric interference with the optimal interaction of the ligand with the biological target.

VI. CHARGE-TRANSFER INTERACTIONS

When a molecule that is a good electron donor comes in contact with a molecule that is a good electron acceptor, the donor may transfer some of its charge to the acceptor.[30,46,47] Such an interaction is named a *charge-transfer interaction*. The charge-transfer complex can form between ions, as shown in Figure 2.3, or between molecules, as shown in Figure 2.4.

The potential energy of a charge-transfer interaction is proportional to the difference between the ionization potential of the donor and the electron affinity of

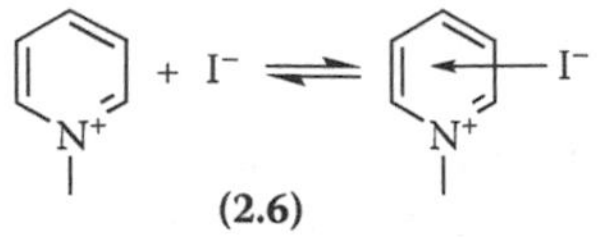

(2.6)

FIGURE 2.3 A diagram of the charge-transfer interaction between the pyridinium and iodide ions.

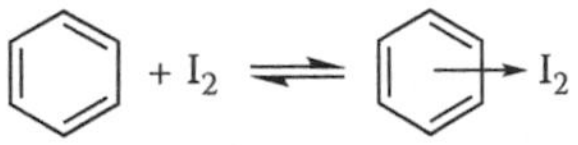

(2.7)

FIGURE 2.4 A diagram of the charge-transfer interaction of benzene and iodine.

the acceptor. That is, the easier it is to remove an electron from the donor or to add one to the acceptor, the greater will be the tendency for such an interaction to occur. The energy of interaction is also dependent on the geometry of the interacting substances; maximum interaction will occur only when there is spatial overlap of the orbitals that are to interact. This can be evaluated with quantum chemical calculations. The ionization potential and electron affinity also correspond to the energies of the highest occupied molecular orbital (HOMO) and lowest unoccupied molecular orbital (LUMO), respectively. Again, these energies may be estimated from quantum chemical calculations.

In terms of substituent constants, ionization potentials and electron affinities within a series may be correlated with Hammett σ values.[48,30] However, there are exceptions; for example, the electron-donating capacity of phenyl carbamates is not completely correlated with σ values or with σ values plus steric substituent constants.[49]

VII. HYDROPHOBIC INTERACTIONS

A *hydrophobic interaction* is considered to have formed if two solute molecules at low concentration interact with one another rather than with water and if the interaction between the solute molecules cannot be explained by hydrogen bonding, or electrostatic or charge-transfer interactions. The measure of hydrophobicity, according to this definition, is thus the relative solubility of the liquid phase of the substance in water if hydrogen bonds and ionic interactions are not present in the liquid phase of the substance. The low solubility of hydrocarbons in water is by definition due to their hydrophobicity. For example, the following equilibrium is a measure of hydrophobic interactions of hexane with hexane:

$$\text{Hexane}_{\text{water solution}} \rightleftharpoons \text{Hexane}_{\text{liquid}} \qquad (2.19)$$

Studies of the solubility of hydrocarbons as a function of structure have provided fundamental information about hydrophobic interactions.[50] This reaction is accompanied by a large decrease in entropy; a hydrocarbon–water system is more ordered than a pure hydrocarbon system. It is now generally accepted that hydrophobic forces are the dominant driving force behind noncovalent intermolecular interactions in aqueous solutions.[30,10,51,52]

The energy of a hydrophobic interaction derives from the unique properties of liquid water. As discussed earlier, in liquid water, hydrogen bonds are continually broken and made, sometimes to different partners, on a nanosecond timescale. However, when a nonpolar molecule is dissolved in water, these rapid fluctuations are inhibited, resulting in a decrease in entropy and hence an increase in the free energy of the system.

The term *hydrophobic bond* is not used because it implies that bond strength is due to the interaction of solute molecules with each other. Indeed, exactly the opposite is true; the hydrophobic interaction results mainly from the tendency of water molecules to associate with each other rather than with nonpolar substances. In other words, the dispersion forces of attraction between the nonpolar molecules make only a minor contribution to the strength of the hydrophobic interaction.[53]

For a homologous series—that is, one in which the members differ only in the length of a carbon chain—hydrophobicity is proportional to the surface area of the alkyl chain.[50,54] This is because the surface area is proportional to the number of water molecules with which the substance interacts. The correlation between hydrophobicity and surface area of an alkyl chain explains why branched chains are less hydrophobic than the corresponding straight chains and also why a methyl group is more hydrophobic than methylene.[55]

For substances that have both a polar and a nonpolar function, the plot of hydrophobicity versus chain length or surface area has a lower intercept, but the same slope, as that of similar graph for hydrocarbons. The methylene group directly attached to the polar or ionic group contributes very little to hydrophobicity, and the next one or two methylenes may contribute less than the normal amount.[50,55] This is because the polar function tends to orient the dipoles of the water toward itself, and the adjacent methylene group is apparently within the influence of this structuring effect that is the opposite to that of a hydrophobic substance. Polar groups also distort the electrons of adjacent carbon atoms and thereby also change their interaction with water.

In conjunction with an extensive investigation of intermolecular interactions,[56] Abraham and his colleagues showed that the partition coefficients of 170 various compounds between cyclohexane and water increases with the volume of the compound and decreases as it forms stronger hydrogen bonds.[33]

Because drugs or pesticides are not normally administered within the cell in which they normally act, passage of these substances from the site of administration to the target must involve passage across cell membranes, which have a hydrophobic core in the lipid regions. If the interaction of a molecule with a biological membrane is thought to be important to its potency, then it may be important to consider hydrophobicity in structure–activity calculations. Overall hydrophobicity of a molecule is considered to be an important determinant of the rate or extent of passage of the substance through membranes.[35–37,39,57] It is also correlated with the ability to disrupt a membrane.[58]

Additionally, hydrophobic interactions are the principal driving force behind the interaction of small molecules with proteins. Although many of the hydrophobic side chains are not on the surface of the protein but rather are found in an internal

$$R_1, R_2$$

HO   O N^+-

(2.8)

hydrophobic core, there are often small hydrophobic patches exposed to solution. Further, sometimes proteins change their conformations to accommodate hydrophobic interactions with small molecules. The hydrophobic region in the binding site of a protein is likely to be of limited size and in close juxtaposition to another type of functional group. Hence, hydrophobic interactions with a protein might occur with only certain types of substances (e.g., anions or cations); it might not reach a plateau with the hydrophobicity of the small molecule, and it might be specific for specific positions on the bound molecule. In the latter case, the correlation of potency with hydrophobicity would be not with the overall hydrophobicity of the molecule, but rather with the hydrophobicity of the substituent at the particular position of binding.

The consideration of hydrophobic interactions of only a part of a molecule also allows one to consider enantiomers in traditional QSAR. One group on the asymmetric carbon atom may be obviously involved in the primary ligand–macromolecule interaction, and a second is typically constant. One would thus consider separately the hydrophobic interactions of the two substituents that have different stereochemical relationships with the common portion of the molecule. Consider the choline esters of glycolic acid (Structure 2.8). The hydroxyl and carbonyl groups are those primarily responsible for the selectivity of the interaction of the ligand with the target biomolecule;[59] hence, substituents R_1 and R_2 interact with different parts of the target. If one of these regions on the target is hydrophobic in nature, only one of each pair of enantiomers may be able to participate in this additional binding. For correlation analysis, the hydrophobic effects of substituents R_1 and R_2 of Structure 2.8 would then be considered separately. Specific examples of such cases are considered in a review.[60]

Hydrophobicity of a substituent is commonly parameterized by its effect on the octanol–water partition coefficient of a molecule.[32] Some have advocated the use of molar volume as a predictor of biological activity. As noted earlier, surface area is more clearly related to hydrophobicity. Empirically, the two are quite highly correlated over limited ranges of chain length. Neither is appropriate by itself for series that contain analogs differing in the number, type, or arrangement of polar groups.

VIII.　STERIC REPULSION

Not all intermolecular forces are attractive in nature: When two molecules approach each other, a minimum allowed distance between the two ultimately would be reached. This distance is the sum of the van der Waals radii of the interacting groups. In other words, the van der Waals radius is a measure of the effective size of an atom in noncovalent interactions. In 3D calculations of molecular structure, steric repulsion is generally considered to decrease by an r^{10} or r^{12} function of distance; it becomes important only at very close approach, but at that point the repulsive energy dominates other factors.

Steric repulsion is an important factor in the determination of the rate of the acid-catalyzed hydrolysis of carboxylate esters; the measurements define the Taft steric parameter, E_s. There is a good correlation between E_s and the van der Waals radius.[61,62] However, it is for steric repulsion that consideration of 3D properties of molecules is important.

IX. LESSONS LEARNED

Table 2.1 summarizes the various types of interactions described in this chapter. Note that Hammett σ constants may be correlated with the strength of dispersion and electrostatic, hydrogen, and charge-transfer interactions. Hence, the mechanism of ligand–macromolecule interaction may not be inferred from a correlation with σ. In a similar manner, the energy of the highest occupied molecular orbital is a factor in dispersion and charge-transfer interactions, as well as the usual electrochemical oxidation potential. Thus, the interpretation of a correlation of potency with this energy level is ambiguous.

TABLE 2.1

Physical Properties Correlated with the Strength of Noncovalent Interactions

Interaction	Distance Dependence	Parameter Used for Substituent Effects	Special Properties
Electrostatic	$r - r^3$	Hammett σ values or charge from molecular orbital calculations	Competition by water may be important
Hydrogen bond	Cut-off at ~3 Å	Same as electrostatic interactions except only for the atom involved in the hydrogen bond	Competition by water is important
Dispersion	r^6	MR × ionization potential or MR alone	Ionization potential is a function of σ or E_{HOMO}; it may be reasonably constant
Hydrophobic	Unknown, but close contact is essential	$\log P$	Entropy gain is the main driving force
Charge transfer	Close contact is essential	Hammett σ values or E_{HOMO} and E_{LUMO} from quantum chemical calculations	Usually, important only in cooperation with other bond types
Steric repulsion	r^{12}	van der Waals radii or E_s	3D QSAR facilitates identification of steric repulsion

REFERENCES

1. Cantor, C. R.; Schimmel, P. R. *Biophysical Chemistry.* W. H. Freeman and Company: San Francisco, CA, 1980, Vol. I.
2. Anslyn, E. V.; Dougherty, D. A. *Modern Physical Organic Chemistry.* University Science Books: Sausalito, CA, 2006.
3. Motiejunas, D.; Wade, R. C. In *Comprehensive Medicinal Chemistry II*; Mason, J. S., Ed. Elsevier: Oxford, 2007, Vol. 4, pp. 193–202.
4. Wallace, A. C.; Laskowski, R. A.; Thornton, J. M. *Protein Eng.* **1995**, *8*, 127–34.
5. Nienaber, V.; Davidson, D.; Edalji, R.; Giranda, V.; Klinghofer, V.; Henkin, J.; Magdalinos, P.; Mantei, R.; Merrick, S.; Severin, J. *Structure* **2003**, *8*, 553–63.
6. Klotz, I. M. *Chemical Thermodynamics.* Prentice-Hall: Englewood Cliffs, NJ, 1950.
7. Prabhu, N. V.; Sharp, K. A. *Annu. Rev. Phys. Chem.* **2005**, *56*, 521–48.
8. Williams, D. H.; Stephens, E.; O'Brien, D. P.; Zhou, M. *Angew. Chem., Int. Ed.* **2004**, *43*, 6596–616.
9. Bojesen, E.; Bojescu, I. N. *J. Phys. Chem.* **1996**, *100*, 17981–5.
10. Blokzijl, W.; Engberts, J. B. F. N. *Angew. Chem., Int. Ed.* **1993**, *32*, 1545–79.
11. Klotz, I. M. *Protein Sci.* **1994**, *2*, 1992–9.
12. Liu, K.; Cruzan, J. D.; Saykally, R. J. *Science* **1996**, *271*, 929–33.
13. Maréchal, Y. *The Hydrogen Bond and the Water Molecule: The Physics and Chemistry of Water, Aqueous and Bio Media*, 1st ed. Elsevier: Amsterdam, 2007.
14. Cho, C. H.; Singh, S.; Robinson, G. W. *Faraday Discuss.* **1996**, *1996*, 19–27.
15. Finney, J. L. *Faraday Discuss.* **1996**, 1–18.
16. Smith, J. D.; Cappa, C. D.; Messer, B. M.; Cohen, R. C.; Saykally, R. J. *Science* **2005**, *308*, 7–9.
17. Smith, J. D.; Cappa, C. D.; Wilson, K. R.; Cohen, R. C.; Geissler, P. L.; Saykally, R. J. *Proc. Natl. Acad. Sci. U. S. A.* **2005**, *102*, 14171–4.
18. Head, G. T.; Johnson, M. E. *Proc. Natl. Acad. Sci. U. S. A.* **2006**, *103*, 7973–7.
19. Jorgensen, W. L.; Tiradorives, J. *Perspect. Drug Discovery Des.* **1995**, *3*, 123–38.
20. Meng, E. C.; Caldwell, J. W.; Kollman, P. A. *J. Phys. Chem.* **1996**, *100*, 2367–71.
21. Li, Z.; Lazaridis, T. *J. Phys. Chem. B* **2006**, *110*, 1464–75.
22. Chervenak, M. C.; Toone, E. J. *J. Am. Chem. Soc.* **1994**, *116*, 10533–9.
23. Gilli, P.; Bertolasi, V.; Ferretti, V.; Gilli, G. *J. Am. Chem. Soc.* **1994**, *116*, 909–15.
24. Taylor, R.; Kennard, O. *Acc. Chem. Res.* **1984**, *17*, 320–6.
25. Allen, F. H.; Bird, C. M.; Rowland, R. S.; Harris, S. E.; Schwalbe, C. H. *Acta Crystallogr., Sect. B: Struct. Sci.* **1995**, *51*, 1068–81.
26. Allen, F. H.; Kennard, O.; Taylor, R. *Acc. Chem. Res.* **1983**, *16*, 146–53.
27. Bürgi, H.-B.; Dunitz, J. D. *Structure Correlation*, 1st ed. VCH Verlagsgesellschaft mbH: Weinheim, 1994, Vols. 1 and 2.
28. Mills, J.; Dean, P. M. *J. Comp-Aid. Mol. Des.* **1996**, *10*, 607–22.
29. Klotz, I. M. *Fed. Proc., Fed. Am. Soc. Exp. Biol.* **1965**, *24*, S24–S33.
30. Jencks, W. P. *Catalysis in Chemistry and Enzymology.* McGraw-Hill: New York, 1969, pp. 340–341.
31. Hine, J.; Mookerjee, P. K. *J. Org. Chem.* **1975**, *40*, 292–8.
32. Hansch, C.; Leo, A. *Exploring QSAR: Fundamentals and Applications in Chemistry and Biology.* American Chemical Society: Washington, DC, 1995.
33. Platts, J. A.; Abraham, M. H.; Butina, D.; Hersey, A. *J. Chem. Inf. Comput. Sci.* **2000**, *40*, 71–80.
34. Abraham, M. H.; Platts, J. A. *J. Org. Chem.* **2001**, *66*, 3484–91.
35. Conradi, R. A.; Hilgers, A. R.; Ho, N. F. H.; Burton, P. S. *Pharm. Res.* **1992**, *9*, 435–9.
36. Chikhale, E. G.; Ng, K. Y.; Burton, P. S.; Borchardt, R. T. *Pharm. Res.* **1994**, *11*, 412–9.

37. Paterson, D. A.; Conradi, R. A.; Hilgers, A. R.; Vidmar, T. J.; Burton, P. S. *Quant. Struct.-Act. Relat.* **1994**, *13*, 4–10.
38. Kubinyi, H.; Martin, Y.; van de Waterbeemd, H.; King, J. *Quant. Struct.-Act. Relat.* **1995**, *14*, 931–8771.
39. Lipinski, C. A.; Lombardo, F.; Dominy, B. W.; Feeney, P. J. *Adv. Drug Delivery Rev.* **1997**, *23*, 3–25.
40. Zhao, Y. H.; Le, J.; Abraham, M. H.; Hersey, A.; Eddershaw, P. J.; Luscombe, C. N.; Boutina, D.; Beck, G.; Sherborne, B.; Cooper, I.; Platts, J. A. *J. Pharm. Sci.* **2001**, *90*, 749–84.
41. Ren, S. J.; Das, A.; Lien, E. J. *J. Drug Target.* **1996**, *4*, 103–7.
42. Abraham, M. H.; Ibrahim, A. *Int. J. Pharm.* **2007**, *329*, 129–34.
43. Zhao, Y. H.; Abraham, M. H.; Ibrahim, A.; Fish, P. V.; Cole, S.; Lewis, M. L.; de, G. M. J.; Reynolds, D. P. *J. Chem. Inf. Model.* **2007**, *47*, 170–5.
44. Agin, D.; Hersh, L.; Holtzman, D. *Proc. Natl. Acad. Sci. U. S. A.* **1965**, *53*, 952–8.
45. PCModels; http://www.daylight.com, Eds. Daylight Chemical Information Systems: Irvine, CA.
46. Kier, L. B. *Molecular Orbital Theory in Drug Research.* Academic Press: New York, 1971.
47. Slifkin, M. A. *Charge Transfer Interactions of Biomolecules.* Academic Press: London, 1971.
48. Kosower, E. M. *Molecular Biochemistry.* McGraw-Hill: New York, 1962.
49. Hetnarski, B.; O'Brien, R. D. *J. Agric. Food. Chem.* **1975**, *23*, 709–13.
50. Tanford, C. *The Hydrophobic Effect: Formation of Micelles and Biological Membranes.* Wiley-Interscience: New York, 1973.
51. Rose, G. D.; Wolfenden, R. *Annu. Rev. Biophys. Biomol. Struct.* **1993**, *22*, 381–415.
52. Levy, Y.; Onuchic, J. N. *Annu. Rev. Biophys. Biomol. Struct.* **2006**, *35*, 389–415.
53. Nicholls, A.; Sharp, K. A.; Honig, B. *Proteins* **1991**, *11*, 281–96.
54. Davis, S. S.; Higuchi, T.; Rytting, J. H. In *Advances in Pharmaceutical Sciences*, Bean, H. S., Beckett, A. H., Carless, J. E., Ed. Academic Press: London, 1974, pp. 73–261.
55. Leo, A. *J. Chem. Rev.* **1993**, *93*, 1281–306.
56. Abraham, M. H.; Ibrahim, A.; Zissimos, A. M.; Zhao, Y. H.; Comer, J.; Reynolds, D. P. *Drug Discovery Today* **2002**, *7*, 1056–63.
57. van de Waterbeemd, H.; Camenisch, G.; Folkers, G.; Raevsky, O. A. *Quant. Struct.-Act. Relat.* **1996**, *15*, 480–90.
58. McKarns, S. C.; Hansch, C.; Caldwell, W. S.; Morgan, W. T.; Moore, S. K.; Doolittle, D. J. *Fundam. Appl. Toxicol.* **1997**, *36*, 62–70.
59. Ariens, E. J. In *Drug Design*; Ariens, E. J., Ed. Academic Press: New York, 1971; Vol. I, pp. 149–57.
60. Martin, Y. C. In *Drug Design*; Ariens, E. J., Ed. Academic Press: New York, 1979, Vol. 8, pp. 1–72.
61. Charton, M. *J. Am. Chem. Soc.* **1969**, *91*, 615–8.
62. Kutter, E.; Hansch, C. *J. Med. Chem.* **1969**, *12*, 647–52.

3 Preparation of 3D Structures of Molecules for 3D QSAR

A three-dimensional quantitative structure–activity relationship (3D QSAR) analysis can provide valuable insights into the basis of ligand structure–activity relationships, not only for ligands in which the bound structure is not known but also for those in which it is known. In fact, even when 3D structures of ligand–macromolecular complexes are available, 3D QSAR often provides more accurate predictions than does a general protein–ligand scoring function because it is adjusted to the specific structures of interest.[1,2]

3D QSAR is not appropriate for every data set. The most suitable data sets are those that contain (1) bound structures of all the molecules with a 3D protein structure that is rather constant within the series, (2) at least one potent conformationally constrained compound and a clear method to superimpose the other ligands on it, or (3) a series of structurally related molecules. The least suitable are data sets for which all of the molecules are extremely flexible, are structurally diverse, and for which there are no 3D structures of representative ligand–macromolecule complexes. The following discussion emphasizes situations in which there is at least some information about the proposed binding conformation. It is assumed that the reader is familiar with 3D molecular modeling.[3,4]

I. PRELIMINARY INSPECTION OF MOLECULES

It is essential to examine the structures of the molecules in the data set before starting serious molecular modeling. In addition, one should verify that all of the relevant structure–activity data has been collected; sometimes there are inactive analogs or a related series of molecules that can be used to test a preliminary model.

The two-dimensional (2D) structures of the molecules should be inspected to see if multiple stereoisomers should be considered when selecting the confirmation for 3D QSAR. It is especially important to examine the structures for the possibility of tautomerization because sometimes the structures of two related molecules are reported in different tautomeric forms. Figure 3.1 shows some of the important types of tautomerization. Note that it changes hydrogen bond donors into acceptors and vice versa, and that it sometimes changes the 3D structure of the molecule. The

31

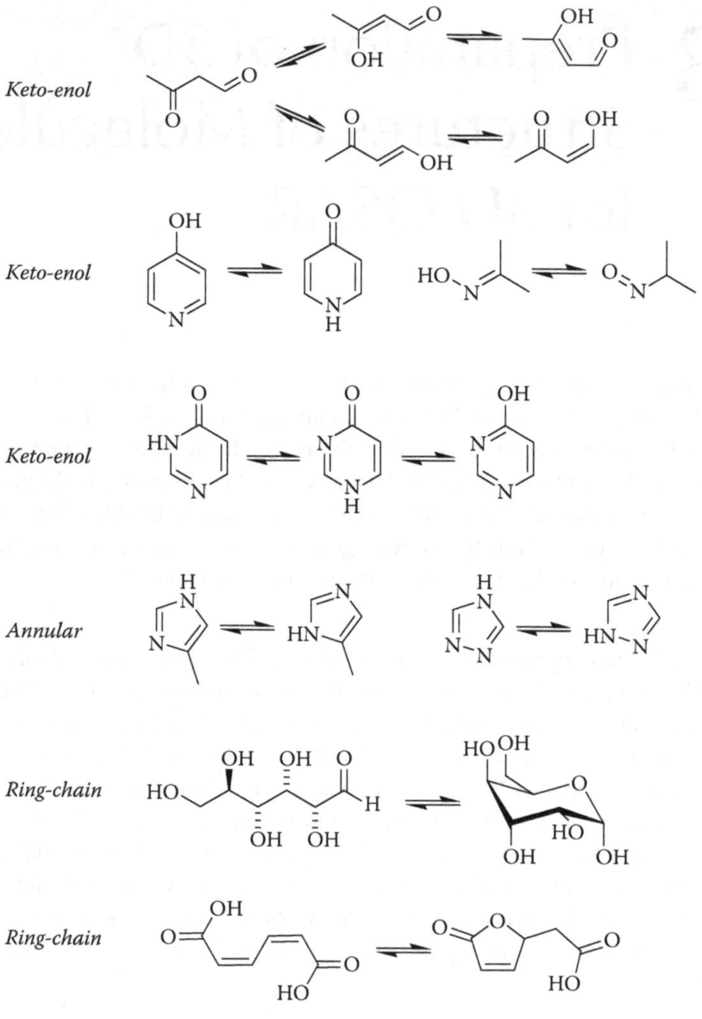

FIGURE 3.1 Examples of common types of tautomerization.

different tautomers of a molecule often have different relative stabilities in different solvents; for example, there can be a different distribution of tautomers in octanol than in water.[5–7] Chapter 6 will discuss the influence of tautomeric stability on the expected relative potency of molecules. There are several programs that generate the tautomers of a molecule.[8–12]

Because 3D QSAR compares the 3D properties of the individual ligands, it is important to use conformations and tautomers of the ligands that emphasize the similarities between them, that is, the 3D arrangement of molecular features associated with biological activity. Typically, these are groups with a specific property such as hydrogen bond donors or those with a positive charge. The choice of features can be known from the observed interactions in the ligand–macromolecule complex or from properties of an active rigid analog. Alternatively, pharmacophore generation software designed for this purpose can suggest the appropriate conformers and tautomers.[13–17] On the other hand, if all molecules of the set have the same possible tautomers and stereoisomers, then consistency is the only option.

Any QSAR, 3D or 2D, is a model that summarizes the available structure–activity data; it is not necessarily a model of reality. A model may be useful and even predictive, but that does not mean that its underlying hypothesis is the truth. Specifically, although there are 3D structures of ligands bound to a particular protein, a more useful model for further analog design may involve changing the conformations of some of the molecules and superimposing the ligands to highlight their similarities and differences. This effort minimizes the subtleties of ligand–protein interactions and also might remedy errors or ambiguities in the experiments.[18–21] Such direct superpositions of ligands also highlight the negative effects of steric bulk at certain positions even though, in fact, the protein may move to accommodate the larger ligand at the cost of binding affinity.

II. GENERATING 3D STRUCTURES OF MOLECULES

A. SOURCES OF STARTING 3D STRUCTURES

Data preparation may start with the 3D structure of a ligand, either bound or free as established experimentally. If no such structure is available, most molecular modeling programs have templates to use for building the molecule of interest. Typically, such templates have been derived from small-molecule x-ray information.[22] Alternatively, there are many programs (e.g., CONCORD,[23] Corina,[24] OMEGA,[25] and Chem3D[26]) that will generate one or more 3D structures of a molecule from a 2D sketch. These programs use templates, energy considerations, and a variety of other strategies to generate the 3D structure.

B. METHODS TO MINIMIZE THE ENERGY OF A 3D MOLECULAR STRUCTURE

To reduce the chance that spurious artifacts will affect the 3D QSAR, the 3D structures generated by various methods often must be energy minimized (optimized) to adjust incorrect bond lengths or angles or, sometimes, to perform slight adjustments to the conformation.

1. Molecular Mechanics

Molecular mechanics[3,4,27,28] is a fast method to optimize a molecular structure; it usually takes only a few seconds for a drug-sized molecule. The method treats the atoms of a molecule as balls and the bonds as springs that hold the molecule together. A molecular mechanics force field uses a set of equations and interdependent parameters to derive energy as a function of a 3D structure. Hence, there are parameters for the distance dependence of the attractive and repulsive energy for each type of atom, and parameters for energy required to stretch and compress the optimum length of each type of bond, to deform a bond away from its ideal shape, and for deviations from an ideal rotation angle. Because each force field uses its own set of parameters that are interdependent, different force fields will optimize the same input structure into a somewhat different 3D structure. For this reason, it is important not to mix-and-match parameters from different force fields.

Force fields are models of reality; hence, it is important to inspect how well a chosen method can optimize the compounds of interest. Usually, this involves a comparison between the structures of one or more observed small-molecule crystal structures[29] and those optimized by the method. For example, some early force fields did not distinguish between the low energy barriers to rotating the single bond between two aromatic atoms, as in biphenyl, and the much higher barrier to rotating the bonds in an aromatic ring out of the plane of the ring. As a result, the programs predicted that biphenyls are planar and the ortho hydrogens are out of the plane of the attached ring.

The advantage of molecular mechanics is that it can be extremely fast in optimizing a structure. The disadvantage is that parameters for unusual functional groups may be missing or guessed incorrectly by the program.

Macromolecular crystallography and nuclear magnetic resonance (NMR) refinements use molecular mechanics to fit the experimental data with a 3D structure that has more-or-less ideal bond lengths and angles. They do so by adding constraints from experimental observations to the other terms in the force field. Because of the experimental constraints and because the QSAR analysis may use a different force field from that used to solve the macromolecule–ligand complex, the ligand structures from such complexes should usually be optimized in the force field that will be used for 3D QSAR.

2. Quantum Chemistry

Quantum chemical calculations discard the notions of bonds, bond angles, and dihedral angles to focus on the interactions of the electrons and nuclei of a molecule. Hence, they can be used to optimize any 3D structure. The only precaution for using these methods is to be sure that the input structure does not have any bad steric contacts. Calculated properties may differ with changes in conformation because the distribution of electrons and nuclei differ with such changes.

Large basis set ab initio or density functional quantum chemistry are the most accurate methods. Because they are orders of magnitude slower than molecular mechanics, they are often used to produce the information necessary to derive parameters for a force field. They are also useful in estimating the relative energies of various tautomers or the relative energies of a hydrogen bond to different atoms of a molecule, and also in calculating the electronic properties of molecules.[30–32]

Semiempirical quantum chemistry programs are faster than ab initio programs because they substitute parameters and assumptions based on experimental data for some of the information needed in a full ab initio calculation. Although somewhat slower than molecular mechanics programs, these methods are useful for calculating electronic properties and for optimizing structures with unusual groups or bonding situations.[3,33,34] These programs are fast enough that they can be used to optimize structures for QSAR applications.

C. METHODS TO SEARCH CONFORMATIONS

If the 3D structure of the molecules bound to the target protein is available, then there is no need to search for other conformations. However, if the bound structures are not available, then conformational searching will probably be required.[35] Although the methods discussed in this section can be used to identify all possible conformations of every ligand in the data set, some of the methods can also be used in a more advanced algorithm to search for conformations that will superimpose key points of multiple ligands.

1. Templates

Starting structures for substituted alicyclic rings can be generated beginning with templates of the ring and adding the substituents in the appropriate stereochemistry. After minimization, the conformations of the side chains can be examined with other methods.

2. Rule-Based Generation of Conformers

There are several programs that generate multiple conformations from one input structure.[25,36–38] They typically use internal templates for the possible rotations about specific types of nonring bonds and for different types of rings. Because they build information into the searching, these programs are quite efficient at searching conformational space for all low-energy conformers. Of course, such methods may suffer from missing or poorly designed parameters. The 3D structures found by these search methods are sometimes further minimized.

3. Rigid Rotation

Conformations can be searched by a systematic rotation about every bond to an sp^3 atom except for symmetrical terminal groups such as methyl or trifluoro methyl. Molecular graphics programs typically support using a dial to perform a rigid rotation around one or two bonds with energies reported as the dial is moved. A more comprehensive rigid rotation would be run in a batch, with the results reported graphically and in a table. The length of the search depends on the number of rotatable bonds, n, and the number of rotation increments, i, by the expression:

$$T = i^n \tag{3.1}$$

For example, one approach is to rotate around sp^3–sp^3 bonds in $120°$ increments and amide bonds in $180°$ increments, and then minimize the resulting structures. Another approach uses very small increments of the rotation, and does not minimize the structures but rather use the energy directly. A typical application might also record distances between key groups as the bonds are rotated.[39–41]

4. Monte Carlo Searching

A Monte Carlo search[3] starts with a minimized 3D structure. At each step of a Monte Carlo search, the algorithm makes a random perturbation of the current structure, minimizes its energy, compares the resulting structure to the conformers that have already been generated, and keeps it if it is unique and lower in relative energy than some prespecified maximum. It also typically keeps some fraction of higher-energy structures using the Metropolis algorithm.[42] It is usual to run a Monte Carlo search until the lowest-energy structure has been sampled typically 10 or 20 times.

5. Molecular Dynamics

Molecular dynamics calculations[43,44] are based on a force field with the assumption that the atoms of the molecule move away from the minimum energy at any temperature above absolute zero,[44] usually, 300K. The movements of the atoms are constrained by force constants relating to bond stretching, angle bending, and torsion rotation, as well as electrostatic and van der Waals attraction and repulsion. As the atoms move, different conformations are sampled. The resulting molecular dynamics trajectory shows how energy and selected dihedral angles change as the molecular dynamics run proceeds. Although it is fascinating to view the changes in structure with time, because it takes many samples to move from a structure at a local energy minimum to another local minimum, molecular dynamics is not the preferred method to search conformational space.

6. Simulated Annealing

Simulated annealing[45] is a special use of molecular dynamics in which a high temperature is maintained for a period and then the system is periodically quenched to optimize the current conformation. This supports a search for the more flexible regions of the molecule but still within the context of the original structure: It is used in crystallographic refinement.[46]

7. Distance Geometry

Distance geometry[47,48] discards the notion of bonds between atoms and instead describes a 3D structure by the distances between all atoms of the structure. It then extends this concept to describe not the distances between atoms in one conformation, but rather the possible minimum and maximum distance between each pair of atoms of the molecule. The minimum distance between any pair of atoms is the sum of their van der Waals' radii. The maximum distance between a pair of bonded atoms is the bond length. The distance between two atoms bonded to a third is a function of the bond lengths and bond angle. The maximum distance between two atoms bonded to the opposite ends of a bond is a function of the respective bond lengths and angles and the maximum rotation angle. The remaining distances are calculated by a method named *bounds smoothing*.

For example, Table 3.1 is the distance bounds matrix that corresponds to ethyl acetate, the extended conformation of which is shown in Figure 3.2. The number above the diagonal is the smallest distance between the indicated atoms and that below the diagonal is the largest distance.

To produce a provisional 3D conformation, the algorithm selects a random distance for each atom pair that is between its minimum and maximum. It then adjusts these distances using the triangle inequality rule, so that no distance between any pair of atoms is larger than the sum of the distances between each of these atoms and every other atom in the structure. Lastly, the algorithm generates a 3D structure by embedding, the atoms by projecting the distances between every atom pair onto three dimensions. The resulting 3D structure is often crude, making refinement necessary.

TABLE 3.1
Smoothed Distance Bounds Matrix for Ethyl Acetate (see Figure 3.2)

	C1	C2	O3	C4	O5	C6	H7	H8	H9	H10	H11	H12	H13	H14
C1	0.00	1.53	2.41	2.58	2.88	3.04	1.09	1.09	1.09	2.15	2.15	2.60	2.60	2.60
C2	1.53	0.00	1.43	2.38	2.64	2.78	2.15	2.15	2.15	1.09	1.09	2.60	2.60	2.60
O3	2.41	1.43	0.00	1.35	2.23	2.48	2.39	2.39	2.39	2.07	2.07	2.55	2.55	2.55
C4	3.68	2.38	1.35	0.00	1.22	1.51	2.61	2.61	2.61	2.43	2.43	2.14	2.14	2.14
O5	4.64	3.51	2.23	1.22	0.00	2.37	2.44	2.44	2.44	2.44	2.44	2.48	2.48	2.48
C6	4.89	3.78	2.48	1.51	2.37	0.00	2.60	2.60	2.60	2.60	2.60	1.09	1.09	1.09
H7	1.09	2.15	3.35	4.77	5.73	5.98	0.00	1.78	1.78	2.25	2.25	2.16	2.16	2.16
H8	1.09	2.15	3.35	4.77	5.73	5.98	1.78	0.00	1.78	2.25	2.25	2.16	2.16	2.16
H9	1.09	2.15	3.35	4.77	5.73	5.98	1.78	1.78	0.00	2.25	2.25	2.16	2.16	2.16
H10	2.15	1.09	2.07	3.29	4.60	4.87	3.05	3.05	3.05	0.00	1.78	2.16	2.16	2.16
H11	2.15	1.09	2.07	3.29	4.60	4.87	3.05	3.05	3.05	1.78	0.00	2.16	2.16	2.16
H12	5.98	4.87	3.36	2.14	3.24	1.09	7.07	7.07	7.07	5.96	5.96	0.00	1.78	1.78
H13	5.98	4.87	3.36	2.14	3.24	1.09	7.07	7.07	7.07	5.96	5.96	1.78	0.00	1.78
H14	5.98	4.87	3.36	2.14	3.24	1.09	7.07	7.07	7.07	5.96	5.96	1.78	1.78	0.00

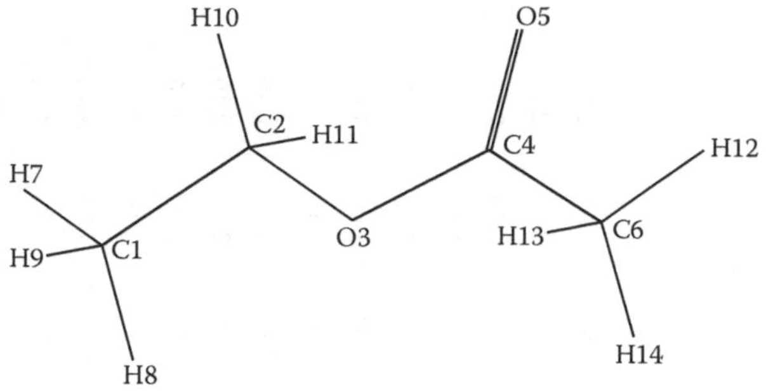

FIGURE 3.2   The 3D structure of ethyl acetate calculated with CONCORD.[23]

We found that distance geometry is an especially useful method to explore the conformations of cyclic compounds for which no preestablished templates are available.[49] Although usually 100 samples are sufficient, provisional 3D structures are generated until the minimum energy structure has been sampled a number of times.

Distance geometry is used to convert NMR distance constraints plus the distance constraints from the covalent structure of the molecule to produce possible 3D structures of the molecule.[50] A similar procedure used the 3D structure of a known protein to model that of a related protein for which the 3D structure was not known. In this case, the 2D structure alignment provided hypotheses as to which distances from the known protein should be preserved in the related protein.[51]

III. STRATEGIES TO SELECT THE CONFORMATION FOR 3D QSAR

Because a QSAR is an abstraction of the structure–activity data, it is not necessary that either the conformations or the implied or actual superposition rules be reflections of reality. Rather, the QSAR should highlight the regions in the ligands postulated to be associated with differences in biological potency.

A. 3D STRUCTURES OF THE LIGAND–BIOMOLECULAR COMPLEX ARE AVAILABLE

X-ray diffraction or neutron-scattering crystallography, NMR experiments, or some sort of docking or molecular simulations may have been used to generate the provisional 3D structures. Although these structures are very informative, they do have limitations.[18–21]

Structures of complexes from both macromolecular NMR and crystallography are models that have been fitted to two pieces of information—the covalent structure of the macromolecule and ligand plus the NMR observables, electron density, or neutron scattering. A force field is used both to regularize the structure and to fit it with the experimental data.

Ligand coordinates may be in error if the force-field parameters are not appropriate for its structure or if there are not enough experimental observables to unambiguously assign the conformation. Furthermore, it is unusual to observe the position of hydrogen atoms in x-ray diffraction; even at resolutions approaching 1 Å, hydrogens might not be visible. A further complication is that at ordinary resolutions many heavy atoms are not distinguishable. As a result, the protonation or tautomeric state of the ligand may be incorrectly assigned and water molecules may be confused with ammonium, chloride, bromide, or sodium ions. For the protein, there might be an error in the rotation of terminal amide or secondary hydroxyl group, or the tautomeric and protonation state of the imidazole of histidine. Even more severe problems may apply to a structure derived from NMR. To use the ligand structures from a complex for a ligand-based 3D QSAR method, additional manipulations of the structures may be necessary. For example, the statistics and interpretability of CoMFA fields are improved if the molecules are superimposed by some common pharmacophoric feature or core rather than by extracting them from superimposed protein structures.[52] It is also usual to optimize the structure, but not change its conformation, to remove artifacts that resulted from the macromolecular structure solution.

Figure 3.3 shows an example of the type of problem that might occur. Figure 3.3a shows the conformation of four analogs as they are bound to the enzyme urokinase. It is clear that three of these analogs share a similar conformation (Figure 3.3b), whereas one was modeled by the crystallographer in a different conformation.[53–55] Because there is no energetic reason why the conformation of the molecule in Figure 3.3c should be different from the other three, we changed the conformation of this molecule to conform to the other analogs and minimized the structures of all four with the Sybyl force field.[56] The superposition of the final structures is shown in Figure 3.3d.

B. 3D STRUCTURES OF SOME OF THE BOUND LIGANDS ARE AVAILABLE

In spite of the limitations discussed in the previous section, the structure of even one ligand–biomolecule complex provides an important starting point for proposing the bioactive conformers, tautomers, and enantiomers of the other ligands. The complex usually identifies the intermolecular interactions involved in binding, thus providing a 3D pharmacophore for superposition. It may also identify parts of the ligand that are exposed to solvent and need not correspond closely in the different molecules.

One may select the conformation, tautomer, enantiomer, and superposition of the other ligands manually with a molecular graphics program. Alternatively, the 3D structure and interactions of the bound ligand could be used to devise a query to search the database of the possible 3D structures of the other ligands[17,57–59] or to generate conformers of the other molecules by adding the pharmacophore constraints as part of a Monte Carlo, distance geometry, or simulated annealing search. If several conformers of a ligand fit the search constraints, the selection between them is usually made by considering their relative energies and how well their volumes overlap with those of the other ligands.

If the 2D structures of some of the ligands vary considerably from that of the bound ligand, the available structures of the complexes may not provide enough information. In such cases, these molecules could be computationally docked into the known macromolecular structure.[60,61] These programs examine the possible orientations and conformations, or poses, of a ligand into a macromolecular binding site. The methods differ in how they compute the relative favorability of the various poses and how they treat molecular flexibility. Some perform rigid docking of the various

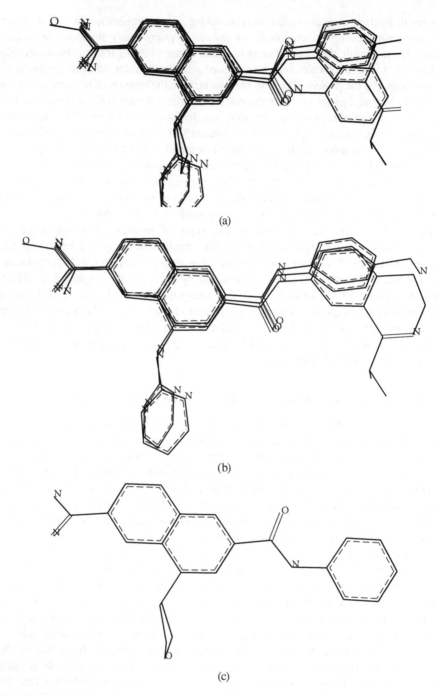

FIGURE 3.3 The conformations of four analogs as they are bound to urokinase[53–55]: (a) all four superimposed using the protein structure, (b) the superposition of three of the analogs that bind in a similar conformation, (c) the 3D structure of the remaining compound, and (d) the four molecules superimposed after conformational and structural minimization.

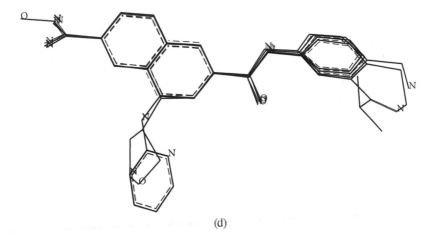

(d)

FIGURE 3.3   (*Continued*)

3D structures of the ligand, whereas others allow either flexibility of the ligand only or flexibility of the ligand and the binding site. The resulting structures of the complexes would then be treated as if they had been generated experimentally.

C. ONE OR MORE CONFORMATIONALLY CONSTRAINED POTENT LIGANDS IS KNOWN

The hypothesis of comparable binding modes of all the compounds that bind to a particular target biomolecule suggests that any conformational constraint in any potent molecule limits the conformations available to the more flexible analogs. Additionally, any structure–activity relationship or bound structure that identifies the groups essential for binding will further constrain the search. Because a rigid analog must bind in its only conformation, this case is similar to that in which the 3D structure of the complex established the bound conformation. There are two missing pieces of information, however: which atoms are involved in specific interactions in the complex with the macromolecular target and which atoms are exposed to the solvent. Hence, the ligand-based procedures discussed in the previous section will be less constrained, with the result that more solutions are possible. Nonetheless, it is possible that one or more tentative models will emerge. These can be tested with subsequent QSAR analysis.

Modeling the bioactive conformation of dopamine at the D2 dopaminergic receptor provides an example of the power of conformationally constrained analogs. Figure 3.4 shows some of the low-energy conformations of dopamine. Several groups of medicinal chemists synthesized compounds in which the catechol amine portion is held in one or another of these conformations.[62,63] They also explored analogs in which one of the phenolic hydroxyl groups is omitted. Figure 3.5 shows the 3D structure and bioactive enantiomer of a molecule that binds to the D2 dopaminergic receptor with high affinity.[64] It also shows the corresponding conformation of the backbone of dopamine. However, the rotation of the hydroxyl group

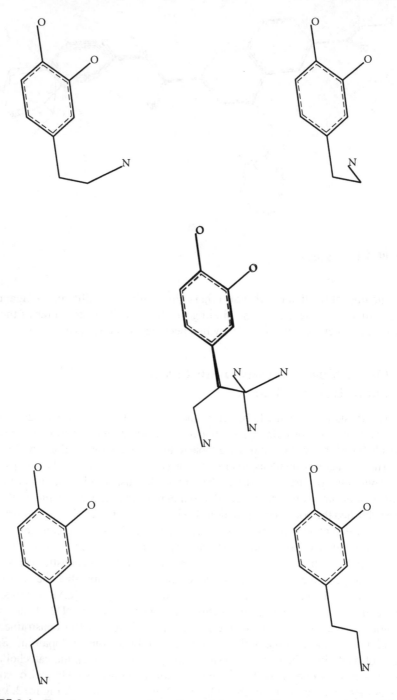

FIGURE 3.4 Four low-energy conformations of dopamine. They are superimposed in the center of the figure.

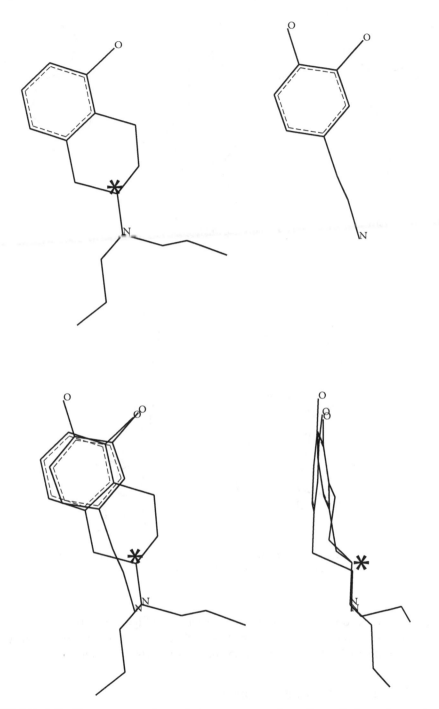

FIGURE 3.5 The correspondence between one conformation of dopamine and a conformationally constrained bioactive phenol. The chiral center of the bioactive enantiomer is indicated with an asterisk.

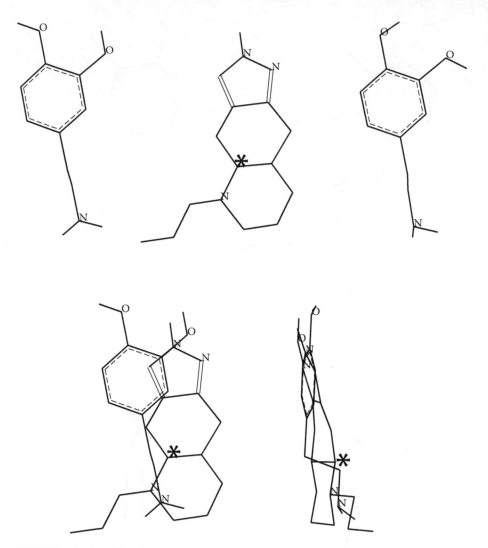

FIGURE 3.6   The use of a heterocycle to establish the hydroxyl rotation of dopamine. The chiral center of the bioactive enantiomer is indicated with an asterisk.

cannot be decided from this information. As shown in Figure 3.6, this was solved by considering heterocycles that also bind to the D2 receptor.[63] The similar orientation of the chiral carbon atom in Figures 3.5 and 3.6 supports these conclusions.

D.   LIGANDS ARE MEMBERS OF A SERIES

Sometimes, the objective of a 3D QSAR is simply to understand the structure–activity relationships of the members of a closely related series. In such cases, usually the common substructure is used to superimpose the ligands. The corresponding conformations of the different molecules would be chosen to maximize the volume overlap with the most potent or largest molecule. The topomer CoMFA method is an example of this approach in which conformations of the variable portions of the molecules are generated automatically.[65]

E. No 3D Structures of Bound Ligands nor Rigid Ligands Are Available

Clearly, this situation provides the most ambiguity and requires the most effort to find satisfactory models. In such a case, some type of automated pharmacophore modeling is required. Pharmacophore modeling typically uses the premise that any constraint on the conformations of any potent ligand also constrains all the other active ligands. It might be that the structure–activity relationships point to a hypothesis as to which atoms should be superimposed in the various compounds: this can simplify the search considerably.

A detailed discussion of methods to detect pharmacophores is beyond the scope of this chapter.[13–17,35] However, some points apply to whatever method is chosen. First, if QSAR is not embedded in the method, it is not necessary to include every active compound; rather, the most constrained from each active series should be considered. Once an algorithm proposes one or more pharmacophores, one can add the remaining molecules using the methods described in Section C. Second, it is not clear what cutoff should be used to discard unreasonably high-energy conformers.[66,67] Third, a pharmacophore is a hypothesis, not a reflection of reality. Even if it leads to predictive 3D QSAR, it is not necessarily a reflection on the actual binding conformation or relative orientation of ligands.

IV. LESSONS LEARNED

Preparing structures of molecules for use in 3D QSAR requires careful consideration of the exact structure that will be modeled. In particular, one must pay attention to the tautomeric state of the molecule and whether the bioactive stereoisomer is known. Although generating 3D conformations is easily accomplished, selecting the conformer for further analysis requires using all possible 3D structural and 2D structure–activity information. Even if a 3D structure of a protein–ligand complex is available, optimizing the structure of the ligand and changing its conformation may be necessary.

REFERENCES

1. Sippl, W.; Holtje, H. D. *J. Mol. Struct.* **2000**, *503*, 31–50.
2. Gohlke, H.; Klebe, G. *J. Med. Chem.* **2002**, *45*, 4153–70.
3. Höltje, H.-D.; Folkers, G. *Molecular Modeling: Basic Principles and Applications*. VCH: Weinheim, 1996, Vol. 5.
4. Leach, A. R. *Molecular Modelling: Principles and Applications*, 2nd ed. Prentice Hall: Englewood Cliffs, NJ, 2001.
5. Plass, M.; Kristl, A.; Abraham, M. H. *J. Chem. Soc., Perkin Trans. 2* **1999**, 2641–6.

6. Abraham, M. H.; Amin, M.; Zissimos, A. M. *Phys. Chem. Chem. Phys.* **2002**, *4*, 5748–52.
7. de Candia, M.; Fossa, P.; Cellamare, A.; Mosti, L.; Carotti, A.; Altomare, C. *Eur. J. Pharm. Sci.* **2005**, *26*, 78–86.
8. ChemAxon Plug-ins, ChemAxon. Budapest. http://www.chemaxon.com/product/calc_pI_land.html.
9. Shelley, J. C.; Cholleti, A.; Frye, L. L.; Greenwood, J. R.; Timlin, M. R.; Uchimaya, M. *J. Comp-Aid. Mol. Des.* **2007**, *21*, 681–91.
10. Tautomer, Daylight Chemical Information Systems, Inc. Aliso Viejo, CA. 2008. http://www.daylight.com/dayhtml/doc/dayprop/index.html.
11. QuacPac, Open Eye Scientific Software, Inc. Santa Fe, NM. 2008. http://www.eyesopen.com/docs/pdf/quacpac.pdf.
12. MOE, Chemical Computing Group. Montreal. 2008. http://www.chemcomp.com/software-chem.htm.
13. Martin, Y. C. In *Designing Bioactive Molecules: Three-Dimensional Techniques and Applications*; Martin, Y. C., Willett, P., Ed. American Chemical Society: Washington, DC, 1998, pp. 121–48.
14. Güner, O. F., Ed. *Pharmacophore Perception, Development, and Use in Drug Design.* International University Line: La Jolla, CA, 1999.
15. Dror, O.; Shulman, P. A.; Nussinov, R.; Wolfson, H., J. *Curr. Med. Chem.* **2004**, *11*, 71–90.
16. Dixon, S. L.; Smondyrev, A. M.; Knoll, E. H.; Rao, S. N.; Shaw, D. E.; Friesner, R. A. *J. Comp-Aid. Mol. Des.* **2006**, *20*, 647–71.
17. Martin, Y. C. In *Comprehensive Medicinal Chemistry II*; Mason, J. S., Ed. Elsevier: Oxford, 2007; Vol. 4, pp. 119–47.
18. Ringe, D.; Petsko, G. A. In *Protein Engineering and Design*; Carey, P. R., Ed. Academic Press: San Diego, 1996, pp. 205–29.
19. Davis, A. M.; Teague, S. J.; Kleywegt, G. J. *Angew. Chem., Int. Ed.* **2003**, *42*, 2718–36.
20. Muchmore, S. W.; Hajduk, P. J. *Curr. Opin. Drug Discovery Dev.* **2003**, *6*, 544–9.
21. Rhodes, G. *Crystallography Made Crystal Clear: A Guide for Users of Macromolecular Models*, 3rd ed. Elsevier: Oxford, 2006.
22. Allen, F., H. *Acta Crystallogr., Sect. A: Found. Crystallogr.* **1998**, *A54*, 758–71.
23. CONCORD, Tripos. St. Louis MO. http://tripos.com/data/SYBYL/Concord_072505.pdf.
24. Corina, Molecular Networks GmbH Computerchemie. Erlangen. www.mol-net.com.
25. OMEGA, OpenEye Scientific Software, 3600 Cerrillos Rd., Suite 1107. Santa Fe NM. www.eyesopen.com/products/applications/omega.html.
26. Chem3D, version 9.0.7. CambridgeSoft. Cambridge, MA. 2005. http://www.cambridge-soft.com/.
27. Boyd, D. B.; Lipkowitz, K. B. *J. Chem. Educ.* **1982**, *59*, 269–74.
28. Burkert, U.; Allinger, N. L. *Molecular Mechanics.* American Chemical Society: Washington, DC, 1982.
29. Allen, F. H. *Acta Crystallogr.* **2002**, *B58*, 380–8.
30. Rablen, P.; Lockman, J.; Jorgensen, W.; Rablen, P. *J. Phys. Chem.* **1998**, *21*, 3782–97.
31. Dimitrova, M.; Ilieva, S.; Galabov, B. *J. Mol. Struct.* **2003**, *657*, 317–24.
32. Jung, D.; Floyd, J.; Gund, T. M. *J. Comput. Chem.* **2004**, *25*, 1385–99.
33. Clark, T. *A Handbook of Computational Chemistry: A Practical Guide to Chemical Structure and Energy Calculations.* Wiley: New York, 1985.
34. Clark, T.; Koch, R. B. *The Chemist's Electronic Book of Orbitals.* Springer: Berlin, 1999.
35. Leach, A. R.; Gillet, V. J. *An Introduction to Chemoinformatics.* Springer: Dordrecht, 2005.
36. Rotate, Molecular Networks. Erlangen. http://www.molecular–networks.com/software/rotate/index.html.
37. Confort, Tripos, Inc. St. Louis, MO. http://tripos.com/data/SYBYL/confort_072505.pdf.

38. Feuston, B. P.; Miller, M. D.; Culberson, J. C.; Nachbar, R. B.; Kearsley, S. K. *J. Chem. Inf. Comput. Sci.* **2001**, *41*, 754–63.
39. Marshall, G. R.; Barry, C. D.; Bosshard, H. E.; Dammkoehler, R. A.; Dunn, D. A. In *Computer-Assisted Drug Design*; Olson, E. C., Christoffersen, R. E., Eds. American Chemical Society: Washington DC, 1979, pp. 205–26.
40. Dammkoehler, R. A.; Karasek, S. F.; Shands, E. F. B.; Marshall, G. R. *J. Comput.-Aided Mol. Des.* **1989**, *3*, 3–21.
41. Beusen, D. D.; Shands, E.; Karasek, S. F.; Marshall, G. R.; Dammkoehler, R. A. *J. Mol. Struct.* **1996**, *370*, 2–3.
42. Metropolis, R.; Rosenbluth, A.; Teller, A.; Teller, E. *J. Chem. Phys.* **1953**, *21*, 1087–92.
43. Rognan, D. In *3D QSAR in Drug Design: Recent Advances*; Kubinyi, H., Folkers, G., Martin, Y. C., Eds; Kluwer: Dordrecht, 1997, pp 181–209.
44. Wassenaar, T. A.; de Vries, A. H.; Mark, A. E., *MD Tutorial: Molecular Dynamics Simulation*, 2005. http://md.chem.rug.nl/education/mdcourse/MDtheory.html (accessed Dec 9, 2008). Molecular Dynamics Group, University of Gronigen. Medium. pdf.
45. Barakat, M. T.; Dean, P. M. *J. Comput.-Aided Mol. Des.* **1990**, *4*, 295–316.
46. Adams, P. D.; Pannu, N. S.; Read, R. J.; Brunger, A. T. *Proc. Natl. Acad. Sci. U. S. A.* **1997**, *94*, 5018–23.
47. Crippen, G. M.; Havel, T. F. *Distance Geometry and Molecular Conformation*. Wiley: New York, 1988.
48. Blaney, J. M.; Dixon, J. S. In *Reviews in Computational Chemistry*; Lipkowitz, K. B., Boyd, D. B., Eds. VCH: New York, 1994, Vol. 5, pp. 299–335.
49. Kim, K. H.; Martin, Y. C. In *QSAR: Quantitative Structure–Activity Relationships in Drug Design*; Fauchere, J. L., Ed. Alan R. Liss: New York, 1989, pp. 325–8.
50. Havel, T. F.; Wüthrich, K. *Bull. Math. Biol.* **1984**, *46*, 673–98.
51. Aszodi, A.; Taylor, W. R. *Fold Des.* **1996**, *1*, 325–34.
52. Klebe, G.; Abraham, U. *J. Med. Chem.* **1993**, *36*, 70–80.
53. Nienaber, V. L.; Richardson, P. L.; Klighofer, V.; Bouska, J. J.; Giranda, V. L.; Greer, J. *Nat. Biotechnol.* **2000**, *18*, 1105–8.
54. Nienaber, V.; Davidson, D.; Edalji, R.; Giranda, V.; Klinghofer, V.; Henkin, J.; Magdalinos, P.; Mantei, R.; Merrick, S.; Severin, J. *Structure* **2003**, *8*, 553–63.
55. Brown, S. P.; Muchmore, S. W. *J. Chem. Inf. Model.* **2007**, *47*, 1493–503.
56. TRIPOS, Inc.: 1699 S. Hanley Road, St Louis, MO 63944.
57. ROCS, OpenEye Scientific Software. Santa Fe, NM. http://www.eyesopen.com/products/applications/rocs.html.
58. Sprague, P. W. *Perspect. Drug Discovery Des.* **1995**, *3*, 1–20.
59. Unity Chemical Information Software, Tripos Associates. St. Louis, MO. 1998. www.tripos.com.
60. Erickson, J. A.; Jalaie, M.; Robertson, D. H.; Lewis, R. A.; Vieth, M. *J. Med. Chem.* **2004**, *47*, 45–55.
61. Warren, G., L.; Andrews, C. W.; Capelli, A. M.; Clarke, B.; La Londe, J.; Lambert, M. H.; Lindvall, M.; Nevins, N.; Semus, S. F.; Senger, S.; Tedesco, G.; Wall, I. D.; Woolven, J. M.; Peishoff, C. E.; Head, M. S. *J. Med. Chem.* **2006**, *49*, 5912–31.
62. Cannon, J. G. *Prog. Drug Res.* **1985**, *29*, 303–414.
63. Seeman, P.; Watanabe, M.; Grigoriadis, D.; Tedesco, J. L.; George, S. R.; Svensson, U.; Lars, J.; Nilsson, G.; Neumeyer, J. L. *Mol. Pharmacol.* **1985**, *28*, 391–9.
64. McDermed, J. D.; Freeman, H. S.; Ferris, R. M. In *Catecholamines: Basic and Clinical Frontiers*; Usdin, E., Ed. Pergamon: New York, 1979, Vol. 1, p. 568.
65. Cramer, R. D. *J. Med. Chem.* **2003**, *46*, 374–88.
66. Ricketts, E. M.; Bradshaw, J.; Hann, M.; Hayes, F.; Tanna, N.; Ricketts, D. M. *J. Chem. Inf. Comput. Sci.* **1993**, *33*, 905–25.
67. Perola, E.; Charifson, P. S. *J. Med. Chem.* **2004**, *47*, 2499–510.

4 Calculating Physical Properties of Molecules

This chapter will discuss how one calculates those properties of molecules that might be involved in ligand–biomolecule binding. The methods attempt to quantitate one or more of the various types of noncovalent interactions. In addition, the discussion will include equations for calculating the relative concentrations of protonated and nonprotonated species from pK_a.

Although this chapter discusses calculating measurable physical properties such as pK_a and octanol–water log P, the reader must remember that the results of these calculations are approximations. Hence, a dose of skepticism should be applied to any calculation, and more than one calculation method should be used whenever possible.

Any error or assumption about the structure of the molecule of interest will be reflected in the results of a calculation using that structure. As discussed in Chapter 3, before performing any calculation of molecular properties, it is prudent to carefully examine the structures of the molecules. For example, if a compound can tautomerize, one must decide whether to include only one tautomer in the analysis—and if so which one—or if several tautomers should be included. Salts also deserve special consideration: Even though one might believe that the ionic form is responsible for activity, ligand-based calculations are usually performed on the corresponding neutral form. This is done because the neutral form more closely resembles the electronics of the bound form, in which any charges on the molecule would be balanced by corresponding charges on the macromolecule. Because some computer programs desalt compounds by removing the smaller component, one should be certain that the fragment of interest is the one used for the calculations.

To calculate molecular properties, one might use the two-dimensional (2D) or three-dimensional (3D) structures of the ligands or the 3D structures of the ligand–macromolecule complex. Of course, combining properties calculated with different methods can also be useful. For example, one might calculate pK_a's of the compounds with substituent values based on their 2D structures and steric properties from their 3D structures. Although proponents of one approach or another might claim otherwise, the two approaches essentially provide alternative views of the same data.[1] The choice of which to use depends on the characteristics of the molecules in the data set and other information that may be available.

If the QSAR involves a set of molecules that contain a common core with variable substituents attached, then the relative properties of the molecules may be found in tables of substituent values.[2] For example, π, σ, and various steric parameters are available.

For sets that are not built around a common core, it is not necessary for QSAR analysis to generate a 3D structure: There are many computer programs that use the 2D structures or SMILES[3] of molecules to calculate the properties[4] or the ability of each ligand to participate in the fundamental intermolecular interactions.[5]

The SMILES language[3,6–8] encodes the 2D chemical structure of a molecule into one line. The SMILES of a structure can be generated from many programs, including the ChemDraw[9] and JME[10] structure editors. These programs, and many others, can convert SMILES to a structure diagram. The appendix summarizes the rules that govern encoding a molecular structure into SMILES.

The physical properties of molecules can also be calculated directly from their 3D structures. The advantage is that, often, more diverse structures can be handled within one analysis.[11–13]

I. THE ELECTRONIC PROPERTIES OF MOLECULES

A. Electronic Properties Calculated from the Structure Diagram

1. σ Values for Substituents on Aromatic Systems

Recall from Chapter 1 that a Hammett σ value is established from the effect of a substituent on the pK_a of benzoic acid (Equation 1.5). The same substituent at different positions of the phenyl ring has different effects on the pK_a and, hence, different σ values. Hansch and Leo provide an excellent review of this topic[14] and tables of values of σ values of many common substituents.[2] An advantage of σ values is that they are derived directly from experimental observations.

One may use the traditional Hammett equation to investigate the structure–activity relationships of other reactions:

$$\log \frac{k_x}{k_o} = \rho\sigma \tag{4.1}$$

The k's are either rate or equilibrium constants; k_o is the constant for the unsubstituted molecule, and k_x is the constant for the substituted one. The ρ value defines the reaction sensitivity to electronic substituent effects: It is defined as 1.00 for ionization of benzoic acids in water at 25°C. The ρ value of some other reactions[15] is shown

TABLE 4.1
Dependence of ρ on Side-Chain and Solvent for Reactions of Carboxylic Acids[15]

Series	Substrate	Solvent	Temp. (°C)	ρ
		Ionization of Acid		
1	ArC(=O)OH	H_2O	25	1.00
2	ArCH$_2$C(=O)OH	H_2O	25	0.49
3	ArCH=CHC(=O)OH(*trans*)	H_2O	25	0.47
4	ArSCH$_2$C(=O)OH	H_2O	25	0.30
5	ArSO$_2$CH$_2$C(=O)OH	H_2O	25	0.25
6	Ar(CH$_2$)$_2$C(=O)OH	H_2O	25	0.21
		Reaction of Acid with Diazodiphenylmethane		
7	ArC(=O)OH	*tert*-Butyl alcohol	30	1.28
8	ArC(=O)OH	Isopropyl alcohol	30	1.07
9	ArC(=O)OH	Ethanol	30	0.94
10	ArC(=O)OH	Methanol	30	0.88
11	ArCH$_2$C(=O)OH	Ethanol	30	0.40
12	Ar(CH$_2$)$_2$C(=O)OH	Ethanol	30	0.22

in Table 4.1. Series 1–6 shows that ρ decreases as more methylenes are inserted between the aromatic ring and the reaction center and that -CH=CH- transmits as much electronic effect as a -CH$_2$-. Series 7–10 shows that ρ decreases as the polarity of the solvent increases. Also, note the similarity of ρ values for Series 1 and 9, for Series 2 and 11, and for Series 6 and 12: Each pair has the same parent structure, but a different equilibrium or reaction is monitored. The value of ρ is not always constant within a structure–activity series: A change in the slope of log k versus σ suggests a change in the mechanism or rate-limiting step.

The σ values derived from the pK_a of benzoic acids do not apply to all reactions; rather, certain functional groups and/or parent molecules and/or reactions require the use of special σ values that include the effect of direct resonance interaction between the substituent and site of reaction. The classic example is the stabilization of the incipient phenolate ion by the *para*-nitro group, Structure 4.1. This resonance stabilization does not occur in *para*-nitro benzoate, Structure 4.2; or in *meta*-nitro phenol.

Therefore, different σ values are used when there is direct resonance interaction between the substituent and the reaction center. If the reaction generates a positive

(4.1)

(4.2)

center adjacent to an aromatic ring, σ^- values are used for the para substituents. The most marked difference between σ and σ^- is for resonance electron withdrawing substituents such as -NO$_2$, -CN, -CO$_2$H, and -SO$_2$NH$_2$. Conversely, if the reaction generates a negative center adjacent to an aromatic ring, one uses σ^+ values for the para substituents. Table 4.2 contains a list of σ, σ^-, and σ^+ values for common substituents selected from the compilation by Hansch and Leo.[2]

2. Separate Field and Resonance Substituent Constants

Physical organic chemists have derived separate σ constants for the resonance and inductive-field effects of substituents: σ_I (inductive) and σ_R (resonance) constants.[16] Thus, for any type of σ value (σ_m and σ_p are different types, as are σ^- and σ^+), only the coefficients a and b of Equation 4.2 differ

$$\sigma = a\sigma_I + b\sigma_R \qquad (4.2)$$

Other workers followed with their own notation, definition, and scales for resonance and inductive effects of substituents.[2,14] A popular one is F and R, field and resonance effects.[17] Typical F and R values are listed in Table 4.2.

TABLE 4.2
Electronic Constants[a] of Common Substituents Sorted by the Value of σ_p

Substituent	σ_m	σ_p	σ_p^+	σ_p^-	F	R	σ_o	σ^*
-N(CH$_3$)$_2$	−0.16	−0.83	−1.70	0.12	0.15	−0.98	−0.36	1.02
-NH$_2$	−0.16	−0.66	−1.30	−0.29	0.08	−0.74	0.03	0.62
-OH	0.12	−0.37	−0.92	−0.37	0.33	−0.70	−0.20	1.37
-OCH$_3$	0.12	−0.27	−0.78	−0.26	0.29	−0.56	0.00	1.77
-Me	−0.07	−0.17	−0.31	−0.17	0.01	−0.18	−0.36	0.00
-OC$_6$H$_5$	0.25	−0.03	−0.50	−0.10	0.37	−0.40	0.67	2.24
-C$_6$H$_5$	0.06	−0.01	−0.18	—	0.12	−0.13	—	0.60
-H	0.00	0.00	0.00	0.00	0.00	0.00	0.00	0.49
-F	0.34	0.06	−0.07	−0.03	0.45	−0.39	0.47	3.19
-Br	0.39	0.23	0.15	0.25	0.45	0.22	0.44	2.80
-Cl	0.37	0.23	0.11	0.19	0.42	−0.19	0.60	2.94
-CONH$_2$	0.28	0.36	—	0.61	0.26	0.10	0.45	1.66
-CO$_2$R	0.36	0.45	0.49	0.75	0.34	0.11	0.51	2.00
-COCH$_3$	0.38	0.50	—	0.84	0.33	0.17	—	1.65
-CF$_3$	0.43	0.54	0.61	0.65	0.38	0.16	—	2.56
-CN	0.56	0.66	0.66	1.00	0.51	0.15	—	3.64
-SO$_2$CH$_3$	0.60	0.72	—	0.82	0.53	0.19	—	3.86
-NO$_2$	0.71	0.78	0.79	1.24	0.65	0.13	1.72	4.66
-N(CH$_3$)$_3^+$	0.88	0.82	0.41	0.77	0.86	−0.04	1.07	4.16

[a] Taken from the tables from Hansch et al.[2] For most σ^- values, there are several values reported. The value reported was that determined under conditions that most resemble the original Hammett definition.

The advantage of separate substituent constants for resonance and field effects is that one does not have to decide which σ value to use. The disadvantage is that two values are needed for each position of substitution. Hence, unless the data set is very large, spurious correlations may be obtained with such a large number of descriptors for each molecule.[18]

Molecules that are substituted at the ortho position present a special problem. From the analysis of 210 sets of data, usually with the unsubstituted analog omitted, Charton concluded that in ortho-substituted molecules there is a delocalized electrical proximity effect.[19] Hence, QSAR analysis of a series that includes ortho-substituted derivatives would require using separate descriptors for inductive and resonance effects at each position of substitution and, because the *ortho*-H does not fit Charton's equation, a steric substituent constant might also be required. Alternatively, typical σ_0 values for ortho substituents in anilines are listed in Table 4.2. These can be used for anilines, phenols, benzimidazoles, etc.

Pyridines can be included in a series that contains substituted phenyl analogs by using the σ values for substitution of the α, β, or δ -CH= of a phenyl ring by –N = of 0.4, 0.6, and 1.0, respectively.[20] However, tautomerization can complicate analysis of heterocycles, as shown by 2-hydroxypyridine, Structure 4.3, which exists mainly in the amide form in water.

(4.3)

3. σ^* Values for Substituents on Aliphatic Systems

Taft characterized the effect of substituents on the equilibria and reaction rates of aliphatic molecules with the substituent constant σ^*, as shown in Equation 4.3.[21]

$$\sigma_X^* = \frac{\log\left|\frac{k_X}{k_H}\right|_b - \log\left|\frac{k_X}{k_H}\right|_a}{2.48} \qquad (4.3)$$

In this equation, k_X is the rate constant for hydrolysis of an ester XCH_2CO_2R', k_H is the rate constant for hydrolysis of the corresponding acetate ester CH_3CO_2R', and the subscripts a and b refer to acid- and base-catalyzed reactions, respectively. The factor 2.48 was chosen so that σ^* would be on approximately the same scale as σ. The term for acid-catalyzed hydrolysis is assumed to be a correction for steric effects on the reaction.[14]

The pK_a's of amines are accurately predicted with σ^* values.[22–24] There are separate equations for primary, secondary, and tertiary amines. The excellent fits of these pK_a's with values derived from the hydrolysis of esters supports the generality of substituent effects in aliphatic systems.

σ^* is a linear function of F and R[17]; hence, one may choose to use F and R values as a measure of electronic properties on nonaromatic systems rather than σ^*.

TABLE 4.3
Examples of Properties Correlated with a Hammett σ Value[15,26]

Equilibria	Rates
pK_a of carboxylic acids	Saponification of esters
pK_a of amines	Esterification of acids
pK_a of benzene boronic acids	Hydrolysis of amides and anhydrides
pK_a of phosphonic acids	Nucleophilic attack of arylhalides
pK_a of phenols	Cleavage of expoxides
pK_a of oximes	Alkylation of amines
Redox potentials	Side-chain bromination of acetophenones
Other	Solvolysis of tri-phenyl methylchlorides
Visible spectra	Beckmann rearrangement
Maximum IR frequency and intensity	Aromatic substitution

The standard deviation for the estimate of σ^* from F and R is +0.16. It contains a larger resonance component that does σ_m. If the data set contains substituents for which a σ^* value is not available, it can be estimated from the observation that the insertion of a methylene group between a substituent and the remainder of the molecule decreases a σ^* value by a factor of 0.36.[25]

Table 4.3 lists examples of properties that are correlated with a σ value.[15,26]

4. Calculations of Partial Atomic Charges, q

Because often in QSAR one does not know which position in the molecules should be used to assign a Hammett σ value, it is helpful to extend the utility of σ values to all positions of a molecule. Calculating the partial charges of each atom would provide the needed information.[27-35] Such empirical charges are typically used both in QSAR,[36] but also as part of 3D modeling. The basis of these calculations is usually a set of parameters derived from an empirical fit of either experimental values such as molecular dipole moments or from fits of electronic properties derived from high-quality quantum chemistry calculations on model molecules.[37] Different programs produce quite different charge distributions because there is no physical basis for deciding how to divide the electrons shared by two atoms. Although the various estimation methods produce reliable trends, charges for a QSAR should all be calculated with the same program.

B. Electronic Properties Calculated from the 3D Structure of the Molecules

Quantum chemical and force-field calculations require using 3D structures to calculate electronic properties of molecules. As noted earlier, from such calculations, one can investigate the electronic effects at all positions of the molecules. Additionally, it is possible to provide estimates of the various σ values from quantum chemical calculations.[38]

Although the postulate of a partial charge q centered on an atomic nucleus is an oversimplification, this simplification is sufficient for many QSAR applications. This may occur because, as discussed in Chapter 2, electrostatic influences on biological potency are attenuated in water. In addition, because in 3D QSAR molecules are generally superimposed so that electrostatically similar groups are overlapped, the result is that the electrostatics is more or less constant within the data set. Lastly, molecules that differ drastically in electrostatic properties from the potent molecules might not bind to the biological target, hence they would not be included in a QSAR analysis.

The most popular representation of 3D electronic properties is the 3D distribution of the electrostatic potential field, the change in potential energy when a charge of +1 is introduced at the points of interest. Figure 4.1 shows the contours of the electrostatic potentials around phenol and imidazole using partial atomic charges calculated with MMFF.[33] Note that the minimum of the potential is in the direction of the lone pairs of the molecule and that the maximum is in the direction of optimal

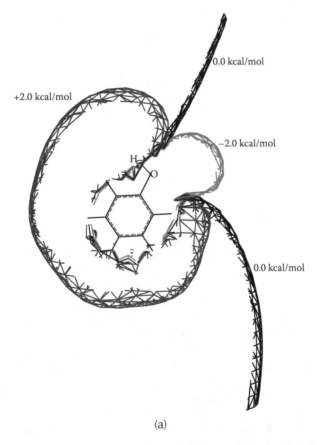

(a)

FIGURE 4.1 A portion of the electrostatic potential around (a) phenol and (b) imidazole calculated with the charges generated with the MMFF94 force field[33] and contoured at −2.0, 0.0, and 2.0 kcal/mol.

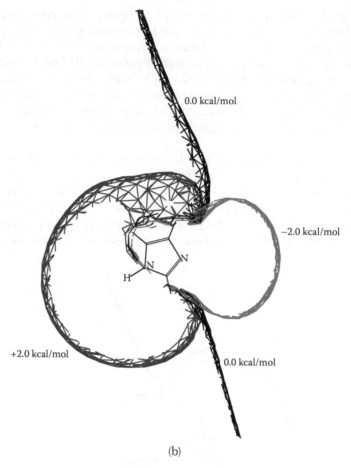

(b)

FIGURE 4.1 *(Continued)*

hydrogen bond donation. The slightly positive contours near the hydrogens attached
to the aromatic carbons are seen in most aromatic systems.

Quantum chemical calculations provide the most detailed picture of the electronic structure of a molecule.[39] We investigated the ability of electrostatic fields to
fit the pK_a values of *meta-* and *para-*substituted benzoic acids that form the basis of
Hammett σ values. We found that with CoMFA, we could reproduce the observed
pK_a's with a standard deviation of 0.072 units when we used the Mulliken[40] charges
from AM1[41] calculations.[38] Charges derived from these semiempirical calculations
were superior to those calculated by fit to atomic charges of the electrostatic potential
from the STO-3G[42,43] wave function and to the charges from the GRID[30] program.
Subsequent studies extended the work to imidazolines and imidazoles[44] and to phenols (unpublished). We used the neutral form of the molecules for the calculations
of the acids and heterocycles. However, because of resonance interaction between
the substituent and the ionizing center for phenols, the fit to σ^- was improved if the
charges were derived from AM1 calculations on the anion.

Others compared several different quantum chemical methods for calculating charges for use in CoMFA and suggest that charges obtained from fitting the semiempirical or *ab initio* wave functions electrostatic potential are superior to charges obtained from empirical methods or Mulliken population analysis.[45] Later investigations showed the superiority HF/3-21G* wave function compared to those from AM1.[46] Others have used a combination of semi-empirical (AM1) and empirically based calculations to generate partial atomic charges that reproduce the results of high-level quantum calculations in a fraction of the computer time.[47,48]

The use of quantum chemical calculations is definitely not limited to estimating atomic charges.[39] They can be used to optimize the geometry of 3D molecular structures and to investigate enzyme mechanisms, the energy of oxidation and reduction reactions, the relative energies of tautomers or conformers, the preferred sites for interaction with a proton or electron, the propensity to participate in charge-transfer interactions, etc. The various molecular descriptors that result from a quantum chemical calculation may also be used as descriptors in a QSAR analysis.

C. pK_a Values and Calculation of Fraction Ionized and Nonionized

1. Sources of pK_a Values

Literature tabulations of pK_a's are found in two chemical society publications,[49,50] in an IUPAC compilation,[51] online,[52] and in the Medchem database.[53]

If one knows the appropriate ρ and σ value and also the pK_a of one member of the series, then Equation 4.1 could be used to calculate the pK_a of another amine,[22-25,54] acid,[55] or heterocycle.[56] As discussed earlier, pK_a's can also be calculated from a 3D QSAR analysis.

There is also software that calculates pK_a values from a computer-drawn structure (Table 4.4).[57-67] Some programs estimate pK_a by recognizing which of hundreds of Hammett equations best describes the input structure, and retrieving or estimating the appropriate σ value. Others are based on other approximations. Most consider tautomerization as part of the calculation. Because of the complexity of the calculation and the possible assumptions that may have been used, it is prudent to validate any calculation on a new molecule with calculations on similar molecules with known measured values.[68-73]

2. Calculation of the Fraction Ionized

If there is variation in the pK_a's of the molecules in the data set of interest, one may wish to examine if the neutral or the ionized form is responsible for the biological activity. How to approach this question will be discussed in more detail in Chapter 6; however, it is essential that one have estimates of the fraction ionized, α, at the pH of interest. The equations for the calculation of α are easily derived from the definition of pK_a:

$$K_a = \frac{[H^+][A^n]}{[HA^{n+1}]} \qquad (4.4)$$

TABLE 4.4

Methods to Calculate pK_a or Octanol–Water Log P from the Molecular Structure

Property Calculated	Program	Type of Log P Calculation	Web Page of Source
pK_a and log P	ADME Boxes	Fragment[63]	pharma-algorithms.com/adme_boxes.htm
pK_a and log P	ACD/pK_a DB and ACD/log P DB	Fragment[62]	www.acdlabs.com/products/ phys_chem_lab/
pK_a and log P	q-Mol	Quantum chemical calculation	www.q-lead.com/qmol
pK_a and log P	Marvin add-in	Atom based[58]	www.chemaxon.com/marvin/
pK_a and log P	QikProp and Epik	3D properties plus a few fragments[64,67]	www.schrodinger.com/
pK_a and log P	PrologP	Neural network on fragments plus Rekker f values plus ALOGP[59,60,61,73]	www.compudrug.com/
Log P	CLOGP	Fragment[69,70]	biobyte.com/
Log P	KowWin	Fragment[71]	esc.syrres.com/interkow/webprop.exe
Log P	XLOGP	Atom-based[72]	inka.mssm.edu/docs/molmod/xlogp/
Log P	ALOGP	Atom-based[68]	www.vcclab.org/lab/alogps/
Log P	miLogP2.2	Fragment	www.molinspiration.com/services
Log P	ChemDraw	Fragment plus CLOGP	www.cambridgesoft.com/software/ ChemOffice/ChemDraw
pK_a	MoKa	Equations based on 3D fields[66]	www.moldiscovery.com
pK_a	SPARC	Quantum chemistry[59]	ibmlc2.chem.uga.edu/spare

In the preceding equation, [H⁺] is the hydrogen ion concentration of the solution, [Aⁿ] is the concentration of the nonprotonated form of the acid, and [HAⁿ⁺¹] is the concentration of the protonated form.

If the nonprotonated species is the ionic one, a carboxylic acid, for example, then α is calculated:

$$\alpha = \frac{[A^-]}{[HA]+[A^-]} \tag{4.5}$$

Dividing each term by [A⁻] and substituting in Equation 4.4:

$$\alpha = \frac{1}{\frac{[HA]}{[A^-]}+1} = \frac{1}{\frac{[H^+]}{K_a}+1} = \frac{1}{10^{pK_a-pH}+1} \tag{4.6}$$

The fraction not ionized, $1-\alpha$, is calculated in a similar fashion:

$$1-\alpha = \frac{1}{1+\frac{K_a}{[H^+]}} = \frac{1}{10^{pH-pK_a}+1} \tag{4.7}$$

If the protonated species is the ionic one, an amine, for example, then the equations are

$$\alpha = \frac{[HA^+]}{[A]+[HA^+]} = \frac{1}{10^{pH-pK_a}+1} \tag{4.8}$$

and

$$1-\alpha = \frac{1}{1+\dfrac{[H^+]}{K_a}} = \frac{1}{10^{pK_a-pH}+1} \tag{4.9}$$

If the molecule is dibasic and the mono-protonated form is the neutral species:

$$1-\alpha = \frac{[HA]}{[H_2A^+]+[HA]+[A^-]} = \frac{1}{1+10^{pK_1-pH}+10^{pH-pK_2}} \tag{4.10}$$

K_1 refers to the equilibrium constant for removal of a proton from the cation to form the neutral species and K_2 to the removal of a proton from the neutral species to form the anion.

From Equation 4.7, one may derive the relationship between pH, pK_a, and the neutral fraction of a molecule that ionizes as an anion:

$$\log(1-\alpha) = -\log(1+10^{pH-pK_a}) \tag{4.11}$$

Figure 4.2 is a plot of log $(1 - \alpha)$ versus (pH $-$ pK_a) for this equation; that is, for a molecule in which the protonated form is neutral and the nonprotonated form is ionized. For molecules in which the protonated form is the ion, the graph would be identical except that the x-axis would be pK_a $-$ pH.

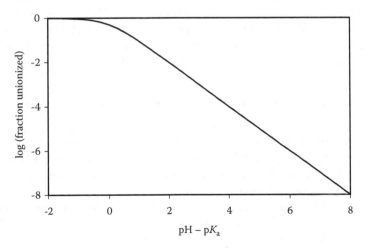

FIGURE 4.2 A plot of log $(1 - \alpha)$ versus (pH $-$ pK_a) for molecules for which the unprotonated form is an anion (Equation 4.11).

It can be seen that for pHs one unit or more below the pK_a, that is, in the direction of decreasing ionization, the fraction not ionized does not change substantially with variations in pK_a. For example, when the pH of the solution is 2.0 units below the pK_a of the molecule, 99% of it is in the neutral form and even at a pH 1.0 unit below, 90% is in the neutral form. Equation 4.7 also simplifies when the pH is above the pK_a. In this instance, $(K_a/[H^+])$ is much greater than 1.0. The result is that

$$\log(1-\alpha) = pK_a - pH \tag{4.12}$$

Hence, if the pH of testing is constant and at least 0.5 log units away from the pK_a to the side favoring the ion, then $\log(1-\alpha)$ is proportional to pK_a. In series that contain such molecules, the regression analysis should include examination of pK_a or some function of it as a molecular descriptor. If the series also contains analog for which the pK_a and pH are nearly equal, then the explicit value indicated in Equation 4.7 or 4.9 must be used.

II. THE HYDROGEN-BONDING PROPERTIES OF MOLECULES

A. COUNTS OF HYDROGEN BOND DONORS AND ACCEPTORS

Frequently, the hydrogen-bonding properties of molecules are tabulated as simple counts: Hydrogen bond donors would be the total number of -NH, $-NH_2$, $-NH_3$, and -OH groups.[74] Sometimes, -SH is also included. For hydrogen bond acceptors, the sum of the number of oxygen and nitrogen atoms would be used.[74]

B. POLAR SURFACE AREA

Polar surface area (PSA), is also a property in which all potential hydrogen-bonding atoms are considered equal: PSA is calculated as the molecular surface area that is closest to an oxygen, a nitrogen, or a hydrogen attached to an oxygen or nitrogen.[75] Originally, PSA was calculated from 3D structures. The exact value of the PSA is relatively independent of conformation unless the conformers differ in the number of internal hydrogen bonds. However, PSA does depend on the atomic radii used to define the surface. More recently, a method to calculate PSA from 2D structures, TPSA, has become available.[76]

C. HYDROGEN-BONDING QUANTITATION BASED ON 2D STRUCTURES

As discussed in Chapter 2, Abraham and colleagues have derived experimentally based quantitative measures of hydrogen bond donor and hydrogen bond acceptor affinities of functional groups, hydrogen bond acidity, and hydrogen bond basicity in their nomenclature.[77] A computer program is available for the calculations.[78] Typical values are listed in Tables 4.5 and 4.6.

Raevsky and colleagues measured the strength of hundreds of hydrogen-bonding complexes to propose substituent constants for hydrogen bond donating and accepting ability.[79] They also offer a computer program.[80]

Because Hammett σ values describe the effects of substituents on pK_a, the full transfer of a proton from one group to another, Hammett σ values also quantitate the

TABLE 4.5
Hydrogen-Bonding Values for Aliphatic Substituents

Substituent	Effect on the Energy to Form a 1:1 Complex		Effect on the Energy to Form the Maximum Interaction	
	Donor α_2^H	Acceptor β_2^H	Donor (sA)	Acceptor (sB)
-H	0	0	0	0
-Cl	0	0.15	0	0.10
-CN	0	0.44	0	0.36
-NH$_2$	0	0.70	0.16	0.61
-NHEt	0	0.70	0.08	0.69
-NEt$_2$	0	0.67	0	0.79
-NO$_2$	0	0.25	0	0.31
-OH (secondary)	0.32	0.47	0.33	0.56
-OEt	0	0.46	0	0.45
-P(=O)(OMe)$_2$	0	0.81	0	1.08
-C(=O)Et	0	0.48	0	0.51
-C(=O)OH	0.54	0.42	0.6	0.45
-C(=O)OEt	0	0.45	0	0.45
-C(=O)NH$_2$	—	—	0.55	0.68
-C(=O)NHMe	0.38	0.72	0.4	0.71
-C(=O)NMe$_2$	0	0.75	0	0.78
-SH	0	0.16	0	0.24
-SEt	0	0.29	0	0.32
-S(=O)Me	0	0.78	0	0.97
Me-S(=O)NMe$_2$	0	0.74	0	—

Source: Used with permission from Abraham, M. H.; Platts, J. A. *J. Org. Chem.* **2001**, *66*, 3484–91.

strength of hydrogen bonds, which are partial transfers of a proton from one group to another.

D. HYDROGEN-BONDING QUANTITATION BASED ON 3D STRUCTURES

For most purposes, the simple electrostatic interaction with a proton is sufficient.[81] Satisfactory accuracy is achieved with semiempirical programs, for example, AM1.

Alternatively, one may calculate interaction energies or electrostatic potentials with a force field using a probe atom that has hydrogen-bonding properties.[27,82–85] In molecular mechanics programs, the electrostatic component of a hydrogen bond is accounted for in the charge calculations from the partial atomic charges of the interacting atoms.[33,86] Sometimes, the radius on the hydrogen atom is set to zero to allow a close approach of an electronegative atom. More accurate results may be found if the partial atomic charges are calculated with quantum mechanics rather than an empirical equation.[38,44] Other programs include a term that penalizes deviation of the hydrogen bond from the ideal angle.[30,84,85] Because the parameters in these

TABLE 4.6

Hydrogen-Bonding Values for Aromatic Substituents

Substituent	Effect on the Energy to Form a 1:1 Complex		Effect on the Energy to Form the Maximum Interaction	
	α_2^H	$\Delta\beta_2^H$	sA	sB
-H	0	0	0	0
-Me	0	0	0	0
-OH	0.6	0.08	0.6	0.16
-OMe	0	0.12	0	0.15
-C(=O)Me	0	0.34	0	0.34
-C(=O)OH	0.59	0.31	0.59	0.26
-C(=O)OEt	0	0.28	0	0.32
-C(=O)NH$_2$	—	—	0.49	0.53
-C(=O)NMe$_2$	0	0.53	0	0.84
-CN	0	0.28	0	0.19
-NH$_2$	0.26	0.24	0.26	0.27
-NHMe	0.17	—	0.17	0.29
-NMe$_2$	0	0.21	0	0.27
-NHC(=O)Et	0.48	—	0.46	0.55
-NHC(=O)NH$_2$	—	—	0.77	0.63
-NO$_2$	0	0.2	0	0.14
-SO$_2$NH$_2$	—	—	0.55	0.66
-P(=O)Me$_2$	0	0.78	0	—
-OP(=O)Pr$_2$	0	0.66	0	—

Source: Used with permission from Abraham, M. H.; Platts, J. A. *J. Org. Chem.* **2001**, *66*, 3484–91.

programs balance the partial atomic charges and the geometry of the hydrogen bond, it is wise to be cautious about changing the charges.

III. THE SIZE OF SUBSTITUENTS AND SHAPE OF MOLECULES

The effect of changing molecular structure on the steric properties of molecules presents the greatest problem for estimating values from the structure diagram alone:[87] Steric interactions are not easily translated from one molecule or reaction type to another, and it is not always clear which aspect of steric effects might be important to a biological response. Overcoming the difficulty of describing steric differences between molecules is a principal advantage of 3D QSAR methods.

Independent of changes in hydrophobicity, increasing the size of a group on a ligand will increase affinity if there are (attractive) dispersion interactions between the group and the target biomolecule. As noted in Chapter 2, dispersion interactions are the result of induced dipole–induced dipole interactions between atoms. They represent the weak attractive steric interactions that molecular mechanics calculates with an r^6 dependence on the distance between the interacting atoms. A positive dispersion interaction implies that all of every substituent can fit into the macromolecular binding site. Nonlinear relationships are observed if beyond a certain size the substituent is too large to fit into the binding site.

Increasing the size of a substituent ultimately results in steric interference with binding. Usually, molecular mechanics calculates that such repulsions increase in strength with r^9 to r^{12} dependence on the distance between the interacting atoms. This is the basis of steric hindrance of certain reactions or interactions.

A. CALCULATING STERIC EFFECTS FROM 2D MOLECULAR STRUCTURES OF RELATED MOLECULES

1. Molar Refractivity

Hansch has pioneered the use of the molar refractivity of a substituent as a measure of its size. Molar refractivity is easily measured and calculated. The MR values of common atoms and groups[88] are listed in Table 4.7 and the published substituent constant tables.[2]

TABLE 4.7
MR, π, and f Values for Aliphatic Substituents

Substituent	MR[2]	$\pi_{aliphatic}$[2]	f[88]
-F	0.09	−0.17	−0.213
-H	0.10	0.00	0.2045
>C<	0.26	—	0.110
-OH	0.28	−1.12	−1.448
-CH<	0.36	—	0.315
-CH$_2$-	0.46	—	0.519
-NH$_2$	0.54	−1.19	−1.34
-CH$_3$	0.56	0.50	0.824
-Cl	0.60	0.39	0.276
-CN	0.63	−0.84	−1.031
-NO$_2$	0.74	−0.85	−0.915
-C(=O)N<	0.78	—	−2.859
-C(=O)NH-	0.88	—	−2.435
-Br	0.89	0.60	0.477
-C(=O)NH$_2$	0.98	−1.71	−1.135
-C$_4$H$_4$N (pyrrolyl)	1.95	0.95	0.615
-C$_5$H$_4$ (pyridyl)	2.30	0.46	0.534
-C$_4$H$_3$S (thienyl)	2.40	1.61	1.613
-C$_6$H$_5$ (phenyl)	2.54	2.15	1.903
Corrections for bonds and branching, per bond or branch			
A chain of saturated hydrocarbons			0.438
A ring of saturated hydrocarbons			0.438
Corrections for proximity effects			
Two electronegative substituents attached to the same carbon atom			0.657
Two substituents attached to adjacent carbon atoms			0.438

MR is additive; hence, each methylene group adds a constant amount to it. Because, as will be discussed in the following text, each methylene group also adds a constant amount to octanol–water log P, in a strictly homologous series, log P and MR will be correlated and one will not be able to distinguish between the contribution of hydrophobicity and dispersion interactions to potency. The correlation between MR and log P breaks down if the series contains different types of groups.

Because of the strict additivity of MR, the value is the same for all isomers of a molecule: It contains no shape information, only size.

2. E_s Values

As discussed in Chapter 1, the classic linear free energy measure of the steric effect of a group is the Taft steric substituent constant, E_s.[21] It is defined as the logarithm of the relative rate of the acid-catalyzed hydrolysis of an acyl-substituted ester compared to that of the corresponding acetate ester:

$$E_{s,X} = \log k_{XCH_2CO_2R} - \log k_{CH_3CO_2R} \qquad (1.6)$$

Although the original definition set the methyl group as 0.0, making hydrogen 1.12, Hansch has championed using hydrogen as 0, making methyl −1.12. It is a measure of the steric hindrance of a group to interaction with the reaction center, not of the overall shape of the molecule. This is seen in Table 4.8 of typical E_s values, which shows that the E_s values plateau at n-butyl for n-alkyl substituents. A complete list of E_s values is found in the table of substituent constants.[2]

TABLE 4.8
Examples of Taft E_s Values

Substituent	Structure	Taft E_s Value[2]
Hydrogen	-H	0.00
Cyano	$-C \equiv N$	−0.51
Fluoro	-F	−0.55
Hydroxyl	-OH	−0.55
Chloro	-Cl	−0.97
Methyl	$-CH_3$	−1.12
Bromo	-Br	−1.16
Ethyl	$-CH_2CH_3$	−1.31
n-Propyl	$-CH_2CH_2CH_3$	−1.43
n-Butyl	$-CH_2CH_2CH_2CH_3$	−1.63
Amyl	$-CH_2CH_2CH_2CH_2CH_3$	−1.64
Isopropyl	$-CH(CH_3)_2$	−1.71
Nitro		−2.52
Phenyl		−3.43

(4.4)   **(4.5)**   **(4.6)**

The assumption that the rate of acid-catalyzed hydrolyses are controlled only by steric factors, that electronic factors do not operate, was controversial from the time of its proposal in the 1950s. The principal evidence that electronic effects are absent is that meta- or para-substituents scarcely affect the rates of both acid-catalyzed hydrolysis of benzoate esters and acid-catalyzed esterification. In contrast, such substituents markedly influence the corresponding base-catalyzed reactions of the same molecules. However, a CoMFA analysis revealed that E_s is approximately 10% electrostatic.[89]

The model reaction may be under the influence of hyperconjugation, with the implication that E_s values should be corrected to[90]

$$E_{s,c} = E_s + 0.306(n - 3) \qquad (4.13)$$

where n is the number of α hydrogens.

E_s values of many substituents cannot be measured because the resulting esters would not be stable. For spherically symmetrical groups, there is a good correlation between E_s and the group radius.[91,92] This provides a method for calculating additional E_s values.

E_s values are a measure of substituent size only if the variable substituent is not restricted in its conformation. For example, although there is a linear relationship between E_s and the rate of quaternization of the nitrogen of Structures 4.4 and 4.5, the isopropyl analog of Structure 4.6 is quaternized 10-fold more slowly than predicted from the rates of quaternization of the methyl and *t*-butyl analogs.[93]

3. Sterimol Substituent Values

Sterimol substituent values were devised to provide steric descriptors for any molecule. They are calculated directly from 3D structures of the substituent in an arbitrary conformation.[94] One Sterimol substituent value is L, the length of the substituent along the axis of the bond between the substituent and the parent molecule. Four width substituent values, B_1–B_4, are measured perpendicular to this bond axis. B_1 is the minimal width; B_2, B_3, and B_4 are distances measured perpendicular to all other B values and ordered in ascending order. Hence, L and B_1–B_4 describe the positions, relative to the point of attachment and the bond axis, of five planes that closely surround the group. An additional substituent value B_5 measures the maximum width perpendicular to the attachment bond. Sterimol substituent values for common substituents are also tabulated.[2]

4. Steric Properties as a Component of Composite Properties

Section VI will discuss two other types of molecular descriptors that contain steric and other types of information: molecular connectivity indices and autocorrelation vectors.

B. CALCULATING STERIC EFFECTS FROM 3D MOLECULAR STRUCTURES

Use of a 3D molecular structure is the most direct way to calculate steric properties of molecules. For CoMFA[95] and GRID/Golpe,[96] this is accomplished by calculating the molecular mechanics interaction energy of the molecule with a methyl group at the lattice points that surround each superimposed molecule. Because the exact location of the lattice with respect to the molecules can affect the results, one may choose to calculate the steric energies with several relative lattice orientations.[97,98]

Catalyst[99] and Phase[100] use QSAR to identify the 3D location of properties in space around a molecule that are favored or disfavored for bioactivity.

IV. THE HYDROPHOBIC PROPERTIES OF MOLECULES

Because hydrophobic interactions are frequently the driving force for the interaction between a ligand and a macromolecule, much attention has been paid to the calculation of the molecular hydrophobicity of biologically active molecules.[101]

A. EXPERIMENTAL MEASURES OF LOG D AND LOG P

Because it cannot participate in either electrostatic or hydrogen-bonding interactions, hydrocarbon–water partitioning would be the preferred measurement to establish relative hydrophobicity.[102,103] However, when polar molecules partition into a hydrocarbon, they bring along associated water molecules. Hence, the hydrocarbon phase contains several different species of solute, with the result that the measured partition coefficient cannot be easily interpreted in terms of fundamental molecular interactions.[104] Additionally, most druglike molecules are not soluble enough in a hydrocarbon to enable accurate quantitation of their concentration in that phase.

In the search for a replacement for the hydrocarbon, a number of more polar organic solvents have been used as the nonaqueous phase for hydrophobicity estimation. However, 1-octanol is most widely used[101] for several practical reasons.[104] Druglike compounds are soluble enough in octanol that one can make an accurate measurement of the partition coefficient. Also, in contrast to hydrocarbons, water-saturated octanol has a regular structure that is not changed by addition of solute;[104] it is chemically stable, commercially available, nonvolatile, and does not absorb ultraviolet light. Although the octanol–water log P includes hydrogen-bonding effects in addition to hydrophobicity,[105,106] the practical considerations outweigh the theoretical. By now, thousands of equations that correlate biological potency with the octanol–water log P have been reported.[107] Thus, the utility of octanol as a model solvent for biological partitioning may be accepted; its ultimate utility in understanding hydrophobicity will also depend on additional research.

Because hydrophobicity depends so heavily on the structure of water, it is not surprising to learn that there are linear correlations between the log P's of compounds between various solvent–water systems.[108–110] For compounds that had been partitioned between water and various alcohols, an average correlation coefficient of 0.99 with a standard deviation estimate of 0.132 is observed.[109] For solvents that do not contain a polar group (such as diethyl ether, benzene, cyclohexane, and chloroform), the log P's of hydrogen bond donor and hydrogen bond acceptor solutes are correlated by significantly different equations.[109,111] In fact, the difference can be used to quantitate hydrogen-bonding potential of substituents.[105] The variable contribution of hydrogen bonding to measured solvent–water partition coefficients must be remembered as one interprets QSAR equations.

Measuring octanol–water partition coefficients is in principle very simple. One equilibrates the compound between an octanol-saturated buffer solution and water-saturated octanol and measures the concentration of solute in the two phases. If possible, the pH of the aqueous phase is adjusted so that the compound is in the uncharged form: 0.1 N HCl for acids and 0.1 N NaOH for bases. Additionally, one must be certain that pure substances are used for solute, solvents, and buffers; that equilibrium has been obtained; that the final result is for the equilibrium of the nonionized monomeric species between the two phases; and that only one tautomer is present. Further details are presented by Hansch and Leo.[112] Partition coefficients may also be measured by titration of the pK_a in the presence of two immiscible solvents.[113]

Experimentally, one measures log D, the observed equilibrium ratio of all molecular species between the nonpolar phase and water (Equation 4.14). Equation 4.17 shows the relationship between log P, Equation 4.15, and log D, assuming that the molecule does not ionize in octanol.

$$\log D = \log \frac{[A]_{oct}}{[A]_{aq} + [HA^+]_{aq}} = \log[A]_{oct} - \log([A]_{aq} + [HA^+]_{aq}) \qquad (4.14)$$

$$\log P = \log \frac{[A]_{oct}}{[A]_{aq}} = \log[A]_{oct} - \log[A]_{aq} \qquad (4.15)$$

From Equations 4.14 and 4.15

$$\log[A]_{oct} = \log D + \log([A]_{aq} + [HA^+]_{aq}) = \log P + \log[A]_{aq} \qquad (4.16)$$

and

$$\log D = \log P + \log \frac{[A]_{aq}}{[A]_{aq} + [HA^+]_{aq}} = \log P + \log(1 - \alpha) \qquad (4.17)$$

Chapter 6 will discuss how to independently incorporate ionization and hydrophobicity into QSAR.

Recently, several workers have published high-pressure liquid chromatographic procedures for measuring log P's.[114–120] Such measurements are faster to make than traditional partition coefficients. Although such measurements would be useful for a series of related molecules, the correspondence between partition coefficients determined in this manner and those determined with the octanol–water system remains to be established.[110,121,122]

B. Tables of Measured Log *P* Values

The Master File of measured solvent–water log P values on over 16,000 compounds includes both a comprehensive literature survey and unpublished values from their and others' work.[53] The computer software (also published in book form in 1995[2]) supports searching for information on a specific molecule using a common name, CAS number, or SMILES, and it also supports substructure and similarity searching to find properties of related molecules. A subset of 10,000 of the octanol–water log P measurements, those considered to be especially reliable, forms the Starlist. The database also contains measured aqueous pK_a values.

Syracuse Research Corporation has an online database of many measured physical properties, including log P and pK_a.[52] The database is searchable by CAS number and indirectly by name. The database is also included with its online log P calculator.[123] If a measured value is available, it will be presented in addition to the calculated one.

C. CALCULATION OF OCTANOL–WATER LOG P FROM MOLECULAR STRUCTURE

If the octanol–water log P of a molecule has not been measured, frequently a reliable estimate of it can be calculated using one of many computer programs available for this purpose.[124–126] Examples of commonly available programs are listed in Table 4.4. Most of these programs calculate log P from the 2D structures of the molecules, using equations based on the types of fragments or atoms in the molecule; however, QikProp uses both 3D properties of the structure and a few fragment types.[64]

In some series, it will not be possible to calculate the log P's of the analogs because that of the parent has not been measured. However, one may still use the difference in log P's of the analogs as properties for the structure–activity calculations.

1. Monosubstituted Benzene Derivatives: Definition of π

The experimental basis for optimism that one can calculate log P from a structure is the early observation that a substituent produces a rather constant change in log P when added to different parent molecules. This led to the definition of π, by analogy to σ, as the change in octanol–water log P brought about by substitution of a hydrogen atom of an unsubstituted molecule (YH) with a substituent X:[127]

$$\pi_x = \log P_{YX} - \log P_{YH} \tag{1.4}$$

In general, for substituents on an aromatic ring, π values are rather constant for parent molecules in which there is not a strong hydrophilic group attached directly to the ring (e.g., when the parent is benzene, benzoic acid, or phenylacetic acid).

To illustrate the calculation procedures, Figure 4.3 shows two ways to estimate the log P of 3-methylchlorobenzene (Structure 4.7). The measured log P is 3.28.

Again, by analogy with σ values, a second set of π values, named π^-, applies to substituents on phenols and anilines.[128] In these cases, the hydrophobic effect of a group is the net result of its intrinsic hydrophobicity plus its effect on the hydrophobicity of the phenol or aniline, and vice versa. Hence, it is not unexpected that the largest difference (+0.73 log units) between normal π values and π^- ones is for the p-NO_2 substituent, which interacts by direct resonance with -OH or -NH_2 to decrease the ability of these functional groups to interact with water. The p-NO_2 substituent

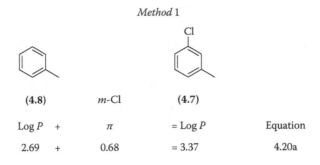

Method 1

(4.8)	*m*-Cl			(4.7)	
Log *P*	+	π		= Log *P*	Equation
2.69	+	0.68		= 3.37	4.20a

Method 2

c1ccccc1
(4.9)   CH₃   *m*-Cl   (4.7)

Log *P*	+	π	+	π		= Log *P*	Equation
2.13	+	0.52	+	0.68		= 3.33	4.20b

FIGURE 4.3 Two ways to calculate the log *P* of Structure 4.7.

increases the log *P* of aniline but decreases that of benzene. Phenoxyacetic acid and benzyl alcohol appear to be intermediate between the two types of parent molecules. Typical values of π and π⁻ are listed in Table 4.9.[88,128] An extensive tabulation is also available.[2]

2. Extension of Octanol–Water Log *P* Calculations to Aliphatic and Complex Systems

Table 4.7 lists examples of π values for substituents in aliphatic systems.

In the early 1970s, Nys and Rekker used regression analysis to establish the average hydrophobicity of -H, -CH<, -CH$_2$-, and -CH$_3$ as well as substituents on saturated carbons.[88,129,130] They also defined a new substituent constant, *f*, as the hydrophobicity of that particular substructure. Log *P* is the sum of constituent *f* values and correction factors *F*:

$$\log P = \sum_n a_n f_n + \sum_m b_m F_m \tag{4.19}$$

in which a_n is the number of occurrences of fragment *f* type *n*, and b_m is the number of occurrences of correction factor *F* type *m*. The current method uses 169 fragment types and 23 correction factors.[131] Table 4.7 lists typical *f* values. Note that with Equation 4.19, log *P* can be calculated even if a value for a parent or reference structure is unavailable. This method is available in the PrologP program, which

TABLE 4.9

π,[128] π,−[128] and f[88] Values for Substituents on an Aromatic Ring Sorted by π_m^-

Group	ortho		meta		para		f
	π_o	π_o^-	π_m	π_m^-	π_p	π_p^-	
-C(=O)NH$_2$	−1.51	−0.57	−1.51	−0.88	—	—	−1.135
-SO$_2$NH$_2$	−2.10	−1.86	−1.50	—	—	—	−1.440
-NH$_2$	−1.40	−0.84	−1.29	−1.29	−1.30	−1.42	−0.902
-SO$_2$CH$_3$	−1.25	−1.20	−1.02	—	—	—	—
-NHC(=O)CH$_3$	−0.14	−0.74	−0.78	−0.73	−0.56	−1.21	—
-OC(=O)CH$_3$	−0.58	−1.02	−0.60	−0.23	−0.58	−1.06	—
-OH	−0.41	−0.58	−0.50	−0.66	−0.61	−0.87	−0.353
-CHO	−0.43	0.24	−0.08	−0.47	−0.06	—	−0.334
-H	0	0	0	0	0	0	0.2045
-N(CH$_3$)$_2$	0.16	−0.48	0.11	0.10	−0.08	−0.69	—
-OCH$_3$	−0.33	−0.13	0.12	0.12	−0.03	−0.12	0.274
-NO$_2$	0.11	0.54	0.22	0.45	—	—	−0.039
-F	0	0.25	0.22	0.47	0.15	0.31	0.444
-CN	0.13	−0.31	0.24	−0.33	0.14	—	−0.155
-CO$_2$CH$_3$	−0.40	0.43	0.50	—	—	—	—
-CH$_3$	0.84	0.49	0.52	0.50	0.60	0.48	0.724
-SCH$_3$	0.30	0.64	0.55	0.87	0.32	—	0.823
-Cl	0.76	0.69	0.77	1.04	0.73	0.93	0.933
-Br	0.84	0.89	0.96	1.17	1.19	1.13	1.134
-I	0.93	1.19	1.18	1.47	1.43	1.45	1.446
-CF$_3$	1.34	1.10	1.49	1.04	1.05	—	1.223
-OC$_6$H$_5$	0.97	0.81	1.56	1.56	1.34	1.46	—
-C$_6$H$_5$	1.92	1.77	1.74	1.74	—	—	1.903

predicts measured values with a mean absolute error of 0.68.[125] It has been extended to alkane–water log P's.

The relationship between π and f values is found by substitution of Equation 4.19 into Equation 1.4:

$$\pi_X = [f_R + f_X] - [f_R + f_H] = f_X - f_H \tag{4.20}$$

Because $f_H = 0.2045$ (Table 4.7):

$$\pi_X = f_X - 0.2045 \tag{4.21}$$

Hence, either the π or the f system may be used to calculate the octanol–water log P. Subsequently, Leo et al. carefully measured the octanol–water log P's of several low-molecular-weight substances in order to establish their own f values.[132] They found that the first single bond of an alkyl group contributes more to log P than do subsequent ones, and that branching decreases log P. The Leo f values are not established statistically, although they are the result of measurements on more than one molecule. The CLOGP program[133] uses the Leo f approach. The 3381 fragment values and correction terms in CLOGP are determined from inspection of the measured log P values of representative compounds.[132,134] CLOGP fits measured values to provide a mean unsigned error of 0.37.[125]

The KowWin program is also based on a regression analysis. It uses approximately 400 descriptors to fit the measured log Ps of 2473 molecules.[71] The program forecasts the log Ps of molecules not used in its training with a mean absolute error of 0.35.[125]

3. Atom-Based Calculations of Log P

In contrast to these fragment schemes for calculation log P, early workers showed that a reasonable fit of log P values of complex substances is descriptors based on atom types alone.[135] Approximately 72% of the variation in log P is explained by a regression analysis using as descriptors the number of nonpolar atoms, the number of nitrogen atoms, the number of oxygen atoms, and correction factors for the proximity of the nitrogens or oxygens. With this method, MLOGP, log P is fitted with a mean absolute standard error of 0.411 and predicted with a mean absolute error of 0.71. This approach to log P calculation is sometimes used in the Lipinski rule of five prediction of permeability.

Ghose and Crippen expanded this approach to include approximately 100 Sybyl[136] atom types that consider not only the atomic number of the atom but also its hybridization and environment.[68,137] Several other groups have used alternative definitions for the atom types.[138,139] Some of these programs include a few corrections for such factors as internal hydrogen bonding. In general, atom-based methods forecast measured log P's with a mean absolute error of approximately 0.60.

4. Incorporating 3D Properties into Calculation of Log P

By incorporating insights gained from decades of 3D modeling of solvation, the QikProp program uses only eight descriptors to fit log P with an error of 0.55: The function includes the molecular volume; the surface area of the weakly polar atoms; counts of the hydrogen bond donors, hydrogen bond acceptors, amines, and amides; and functions of the hydrogen-bonding groups with surface area and dipole moment with volume.[64]

5. Nonadditivity of Substituent Effects on Log P

Several factors contribute to the nonadditivity of fragment contributions to log P. These may or may not have been considered in the computer programs that calculate log P.

An intramolecular hydrogen bond formed in the substituted analog but not the parent can raise the observed log P. For example, one might expect the log P of salicylic acid (Structure 4.10) to be 1.51, based on the log P of phenol (1.47) plus the π of carboxyl (0.04). However, the measured value is 2.26.[2] Because the -OH and

Oc1ccccc1C(=O)O   c1ccccc1OCC(=O)O   Cc1ccccc1OCC(=O)O   c1c(C)cccc1OCC(=O)O

(4.10)   **(4.11)**   **(4.12)**   **(4.13)**

(4.14)

-C(=O)OH of salicylic acid are known to be hydrogen bonded, the increased hydrophobicity is rationalized as being due to an internal hydrogen bond.

A steric effect that prevents the access of water to a hydrophilic group can also lead to nonadditivity of log P. For example, consider Structures 4.11–4.13. An ortho methyl group increases the octanol–water log P of phenoxyacetic acid (Structure 4.11; log P = 1.34) by 0.64 (Structure 4.12; log P = 1.98), but a meta methyl group (Structure 4.13; log P = 1.78) raises it by only 0.44. The latter is more in line with a usual π value of 0.5.[2] Apparently, the ortho methyl group denies water access to the ether oxygen (-O-), as a result of which log P is raised more than expected. This type of interaction results in an even larger deviation from the expected log P when the group that is occluded by the substituent is more polar than -O-.

In contrast, steric interactions can lower log P if polar groups are made more accessible to water. The classic example of this effect is 2,3,4-trimethoxy benzene, for which the observed octanol–water log P is 1.53. However, considering that the log Ps of benzene and methoxy benzene are 2.13 and 2.11, respectively, one would expect that the observed log P would be no lower than 2.0. Apparently, the methoxy groups are more accessible to water in 2,3,4-trimethoxy benzene than in the monosubstituted molecule.

The most accurate manual method—or check of a calculated value—for the calculation of the log P of a complex molecule is to find a measured octanol–water log P of a molecule that differs from the one of interest by the presence or absence of a single well-behaved substituent such as a methylene or a halogen. The calculation of the desired log P then follows either of the preceding sequences depending on whether a π or an f value for the substituent is available.

For example, the log P's of the lincomycin analogs, Structure 4.14, in which R_2 is an alkyl group, may be calculated from the measured octanol–water log P of lincomycin, Structure 4.14, R_2 = n-propyl. Because we were interested in N-methyl analogs that differ from lincomycin only in the number of methylene groups in the substituent at R_2, to calculate the log P, the appropriate number of increments of the f value of methylene were added or subtracted from the log P of lincomycin.

E. CALCULATION OF 3D HYDROPHOBIC FIELDS

For 3D QSAR, we would like to calculate the distribution of hydrophobic fields around the ligands of interest. Typically, for CoMFA the molecules in the set have been superimposed to emphasize similarities in their polar functionality because such groups are responsible for recognition by the biomolecule. In such cases, changes in steric fields might be highly correlated with changes in hydrophobic fields. In other situations, hydrogen bonding or electrostatic fields plus steric fields might be correlated with hydrophobicity.

In contrast, several groups have presented force-field methods to explicitly calculate the 3D hydrophobic fields around molecules. These force fields include a term for the hydrophobicity of a particular atom type and one or more terms for how hydrophobicity depends on the distance between groups in the molecule. They differ in how the parameters for hydrophobicity of individual atoms were chosen and how they treat the distance dependence of the energy of a hydrophobic interaction.

For example, the energy of interactions with the GRID[30] water probe includes not only the usual dispersion and electrostatic attraction and repulsion energies but also terms for deviations (Δd) from the ideal length and angle of the particular type of hydrogen bond as observed in small-molecule crystal structures.[85] Too close an approach is penalized as a function Δd^8, and too distant an approach as a function of Δd^6. The parameters of the earlier version of this probe atom are validated by the observation that hydrophobic fields calculated with it explain the octanol–water log P of a series of analogs.[140] One example of the utility of the GRID force field is its key role in the design of potent neuraminidase inhibitors.[141] It is also the basis of ADME predictions in VolSurf[142] and the molecular descriptors used in the alignment-independent 3D QSAR methods GRIND and ALMOND.[143]

In the HINT (hydrophobic interaction) force field,[144] the hydrophobicity of atoms is parameterized from the CLOGP fragment values, with special consideration given to interaction terms and proximity effects. HINT performs the hydrophobic field calculation using an e^{-d} dependence on the distance d between the probe and the interacting atom. In addition to calculating hydrophobic fields for 3D QSAR, HINT is also useful for docking ligands to a protein[145] and for determining the charge state of protein residues as a function of pH.[146]

In a similar type of hydrophobic force field, the MLP (molecular lipophilicity potential) method[138,147,148] describes each atom using an atomic lipophilicity[149] and an $e^{d/2}$ falloff with the distance. MLP was used in a CoMFA model for nonnucleoside HIV reverse transcriptase inhibitors that showed good predictive ability.[150] It can also be used to calculate log P.[126,151]

F. COMPUTER PROGRAMS THAT CALCULATE OCTANOL–WATER LOG P

The preceding discussion suggests that the many factors that affect an octanol–water log P can make manual calculation problematic. Fortunately, there are many

(4.15)   (4.16)   (4.17)

computer programs that calculate octanol–water log P from structure diagrams or SMILES alone. Table 4.4 lists those that are easily available.

The most complex log P calculator uses fragment values: For example, the CLOGP program uses several thousand fragments in five environments and 56 types of interaction terms. The values of the fragment constants and correction terms in CLOGP are determined from inspection of the measured log P values of representative compounds.[132,134] Several log P calculation programs refine this fragment-based approach by statistically fitting the fragment values and correction terms.[62,126,152,153]

Programs that calculate log P from the atom types only make it possible to calculate the log P of almost any organic molecule. Typically, the relative contribution of approximately 100 atom types is determined by multiple regression analysis of measured log P values.[135,139,154–156]

Table 4.10 shows the log P calculated for some common drugs using different programs.[76,123,133,157–159] It shows that frequently there are substantial differences between the calculated log P values. For example, the relative hydrophobicity of the three beta-lactams (Amoxil, clavulinic acid, and Keflex-Structures 4.15–4.17) is not parallel for the four methods. Also, note the difference in calculated values for the different tautomers of Glucophage and Viagra: the methods do not even agree as to which tautomer is more polar. Of course, measured log P's of molecules that tautomerize are complicated by the fact that both phases might contain several tautomers and that the ratio of tautomers is probably different in the polar and nonpolar phases. These results emphasize that one must be skeptical of any calculation and be alert to issues that may affect its accuracy.

Table 4.10 also shows that the 2D calculations of polar surface area agree quite well with one another but differ from that calculated from 3D. The column for polar surface area calculated to include surface area close to sulfur and phosphorous is included to alert the reader that sometimes polar surface area is defined in this way.

V.   INDICATOR OR SUBSTRUCTURE VARIABLES

A.   INDICATOR VARIABLES COMBINED WITH PHYSICAL PROPERTIES

Frequently, data sets that were not designed for QSAR contain more than one series of molecules. For example, in a set of β-adrenergic blockers, the chain between a substituted aromatic ring and the essential basic nitrogen may be either -OCH$_2$CH(OH)-CH$_2$- or -CH(OH)CH$_2$-. Indicator variables encode such features by using one value to indicate the presence of a particular feature and another to indicate its absence.[160]

TABLE 4.10

Examples of Calculations of Log *P* and PSA on Common Drugs

Name	Structure	Octanol–Water Log *P* Programs				Polar Surface Area Programs			
		CLOGP[133]	KowWin[123]	ALOGP[158]	ACD[159]	Savol 3D[157]	Ertl 2D[76] TPSA	SciTegic 2D[158] TPSA	SciTegic 2D add *S* and *P*[158]
Amoxil	4.15	−1.87	−1.36	−2.54	0.61	155	133	133	158
Keflex	4.16	−1.84	−1.93	−2.26	0.65	129	113	113	138
Clavulanic acid	4.17	−1.06	−2.04	−1.24	−1.98	113	87	87	87
Glucophage	4.18	−1.45	−1.40	−1.39	−1.82	95	91	91	91
Glucophage tautomer 1	4.19	−1.63	−2.64	−0.74	−2.31	95	89	89	89
Glucophage tautomer 2	4.20	−1.63	−1.40	−0.74	−1.82	96	91	91	91
Motrin	4.21	3.68	3.79	3.61	3.72	45	37	37	37
Celebrex	4.22	4.37	3.47	4.43	4.21	91	78	78	86
Lipitor	4.23	4.46	6.36	5.56	4.13	112	118	118	118
Viagra	4.24	2.22	2.30	2.25	2.27	113	113	113	118
Viagra tautomer 1	4.25	3.56	1.60	3.06	3.14	120	114	114	122
Viagra tautomer 2	4.26	1.98	2.47	2.25	2.27	114	113	113	118
Claritin	4.27	5.05	5.66	5.00	5.94	—	42	42	42

(4.18)

(4.19)

(4.20)

(4.21)

(4.22)

(4.23)

(4.24)

(4.25)

(4.26)

(4.27)

One may use 1.0 and 0.0 as these values; in this case, the importance of the feature is easily estimated from the regression equation. One less indicator variable is needed than the number of features to be distinguished. The feature involved may be rather specifically defined, such as *"ortho* methyl," or may be more general, such as "any *ortho* substitution."

Graphically, one may picture an indicator variable as a constant that accounts for the difference in intercepts of two parallel lines. If an indicator variable is to be used, the QSAR for the individual subsets of the data should yield equations in which the same properties are significant and have the same coefficients. In no case should an indicator variable be used for only one molecule; this is equivalent to omitting that molecule from the equation. It is perhaps questionable, although not entirely unreasonable, to use an indicator for only two molecules.

The ultimate in use of indicator variables is the Free–Wilson technique, which is regression analysis that uses only indicator variables to describe the molecules in a series. There are other uses of indicator variables.[160] Using them to combine different biological endpoints is illustrated in Chapter 8, Equation 8.5.

B. SUBSTRUCTURAL DESCRIPTORS AS THE BASIS OF A QSAR

If one wishes to analyze a large data set of diverse structures, traditional physical properties might not produce a significant QSAR. Instead, one can explore if certain 2D molecular features or substructures are associated with a biological property: The hundreds to millions of substructures of possible interest are quickly identified with various computer programs. However, considering such a large number of potential descriptors can lead to spurious results requiring the use of the special algorithms discussed in Chapter 11 as well as rigorous validation techniques such as those discussed in Chapter 7.

One type of computer program identifies only specific predefined substructures: If the substructure is present in a molecule, that element of the output vector is set to 1.0 or to the count of the number of times it occurs; if it is not present, the element is set to 0.0. Examples are the popular MACCS/ISIS keys[161] and the longer, more elaborate substructures encoded in LeadScope.[162]

Other computer programs do not use predefined substructures but rather enumerate them as part of their algorithm. The substructures that the Daylight software enumerates are called *path fingerprints*.[163] They are formed starting from each atom in the structure and consider the character of the atoms in every nonbranched sequence (a path) of bonds. The default range of the number of bonds in a path is from 0 to 7. For example, starting at the oxygen atom of isopropyl alcohol, $HOCH(CH_3)_2$, the following hydrogen-suppressed paths would be identified O, O-C, O-C-C, C, C-C, C-C-C, and C-C-O. In contrast, the substructures that the SciTegic software enumerates are called *circular fingerprints*.[164] They are formed starting at each atom by considering the atoms connected to it, and the atoms connected to them, etc. Usually, they are calculated by considering the environments two or three bonds away from the center, corresponding to path lengths of four or six. For example, starting at the oxygen atom of isopropyl alcohol, the substructures identified would be the same as those identified by the Daylight algorithm, but starting at the branched carbon atom they would also include C(-C)(-C)-O. An alternative formulation of the SciTegic fingerprints identifies the atoms not as their atom types but rather their character: hydrogen bond donor, hydrogen bond acceptor, aromatic, etc.

Such substructure-generating programs typically identify tens of thousands possible substructures within a set of molecules. The raw output would thus be, for every molecule, an array of tens of thousands of items with each position corresponding to a particular substructure. Usually, instead of such an array, the program produces a fingerprint of a predetermined length: Each substructure pattern is used as a seed to a pseudo-random-number generator (it is hashed), to indicate the four or five bits be set in the fingerprint.[163] The final fingerprint includes the contributions of each substructure in the molecule. In this way a large number of substructures are encoded into a shorter bit-string. Although hashing shortens the vector, by doing so it removes the correspondence between any vector element and a particular substructure. In spite of such ambiguity, Chapter 11 will show how SciTegic fingerprints perform in several analyses of the monoamine oxidase data set.

VI. COMPOSITE DESCRIPTORS CALCULATED
FROM 2D STRUCTURES

Some molecular topological indices[165,166] are composite variables that encode both shape and electrostatic properties. The different topological indices are generated from various procedures that consider the properties of the atoms and how they are connected in the 2D molecular structure. For example, one category of topological indices considers the electronic properties of the atoms, such as the number of valence electrons and the electronegativity of the atom as perturbed by other atoms in the structure. Table 4.11 provides a summary of three of the common topological indices. Other variants of this type of index are also calculated.[165,167–170] A caveat is that many such properties can be calculated; hence, if the data set is small, possible spurious correlations may result from a QSAR.[18]

The advantages of topological indices are that they produce predictive QSAR models, they can be calculated for any molecule, and the software is available in many different packages.[170] The main disadvantage is that the models cannot be manually examined to discover what type of features the ideal molecule would have. To get around this problem, the models can be used to predict the potency of all proposed molecules.

VII. PROPERTIES CALCULATED FROM THE 3D STRUCTURE
OF THE LIGAND–MACROMOLECULE COMPLEX

Although a thorough discussion of calculating properties of a ligand–macromolecule complex[171] is beyond the scope of this book, the resulting numbers can be used as input to a QSAR analysis. The resulting QSAR could be thought of as a scoring function derived from a specific set of protein–ligand complexes.[145,172–178] In many of these approaches, it is possible to include different proteins as well as different ligands to produce a quantitative specificity-structure–activity model.

TABLE 4.11
Overview of Some Molecular Connectivity Calculations[170]

Index	How It Is Calculated	Properties Described
Molecular Connectivity Chi: ${}^m\chi_t^{[165,167]}$	The value of a molecular connectivity index depends on the number of times a particular generalized substructure is found in a molecule and the number of sigma electrons in the atoms involved in the individual substructures. The different χ's are distinguished by the value of m, which encodes the number of paths (bonds) in the subgraph (it ranges from zero to four), and by the subscript t, which encodes the type of path such as chain, cycle, or branch.	Size and shape
Kappa Shape: ${}^1\kappa$, ${}^2\kappa$, ${}^3\kappa$ and ${}^1\kappa_\alpha$, ${}^2\kappa_\alpha$, ${}^3\kappa_\alpha$ [165,168]	The *Kappa* values are derived from counts of one-bond, two-bond and three-bond fragments, ${}^1\kappa$, ${}^2\kappa$, ${}^3\kappa$, respectively. Each count is normalized to fragment counts in reference structures that possess a maximum and minimum value for that number of atoms in a molecule. A variation, ${}^m\kappa_\alpha$, also encodes the covalent radius of the atoms.	Size and shape
Electrotopological State (*E*-State): S_i [169]	These numerical values for each atom in a molecule encode information about both the topological environment of that atom and the electronic interactions due to all other atoms in the molecule. A special calculation method is used for hydrogen atoms.	The sum of the hydrogen bond accepting and donating properties; other electrostatic and shape properties.

Before calculating the properties of the complexes, attention must be paid to their structures. For example, one must decide whether or not to use a force field to minimize the energy, and hence change the structures, of the complexes. If it will be minimized, will the whole structure be relaxed, just the ligand, or the ligand plus the active site?[173,179] Will some structures be changed to represent corresponding ligand and protein tautomers or not? Will waters placed by the crystallographer be included in the calculations?[180] How will waters be treated if they are present in some complexes but not others?[181–183] How will the protonation states of ionizable protein groups be treated? What structure will be used if the ligand can bind in more than one pose?

A number of different properties of the complexes could be considered for a QSAR. For example, one might calculate the total interaction energy of the ligand with the complex and investigate its correlation with the observed log $(1/C)$.[173,184,185] Alternatively, one could analyze the correlation of the separate hydrogen-bonding, dispersion, and electrostatic interaction energies of the ligand with the protein or these interaction energies between each protein or water atom in the active site.[182,186] In the latter case, partial least-squares analysis would be used, rather than ordinary least squares.

Knowledge-based approaches to binding affinity prediction are based on the analysis of the structures of known protein–ligand or DNA–ligand complexes.[187–191] The method tabulates the frequency with which each pair of atom types occurs at several distance intervals. This frequency distribution is then converted into an interaction energy equation using the Boltzmann distribution. Hence, knowledge-based potentials ignore the classical components of an energy function and instead directly calculate the free energy of interaction. They perform approximately as well as more traditional scoring methods in docking studies,[192] but this is not precise enough for affinity prediction.

The AFMoC method uses knowledge-based potentials in a CoMFA-like fashion.[177,193,194] It calculates knowledge-based pair-potentials between protein and ligand atoms at intersections of a lattice that bounds the active site. A PLS analysis provides the contribution of each lattice point to affinity.

VIII. ORGANIZING MOLECULAR PROPERTIES FOR 3D QSAR

No matter what the source of properties of molecules, for a 3D QSAR one must generate a set of descriptors whose definition and meaning is constant within the set of molecules, which may be of different size and shape. The objective is to generate a table in which each row represents one molecule and each column represents one property. There may be many properties in the table, but each column must refer to the same property of every molecule.

A. CALCULATIONS BASED ON MOLECULAR SUPERPOSITIONS

A straightforward solution to this problem is to superimpose the molecules and surround them with a lattice. At each intersection point of the lattice, one or more properties is calculated. Thus, a column would correspond to one property calculated in one location in space. This is the data representation that is used in CoMFA,[95] CoMSIA,[195] HASL,[196] and AFMoC.[177,193,194]

In CoMFA, the properties at the lattice points are the energies of interaction with a probe atom, a methyl group in the case of steric energies. Steric energies inside the volume of a molecule will be very large: This is useful in describing negative steric influences on potency. However, they must be truncated at some reasonable value in order to not dominate other effects. Electrostatic energies are calculated only for points outside the molecule.

Although CoMSIA also calculates properties on a lattice, rather than the energy of interaction with a probe, it calculates the similarity of the probe atoms to the fields produced by the molecule. This eliminates the high steric energies associated with probes close to atoms and also permits the calculation of electrostatic energies inside the molecule and 3D hydrophobicity.

HASL[196] uses one molecule to form a reference lattice of points that it occupies, and iteratively superimposes the lattices of other molecules upon it to maximize the number of lattice points in common. The relative shape of two molecules is thus the number of lattice points in common.

One can also use the individual superimposed molecules to calculate the properties at points other than on a lattice. For example, the COMPASS[197] method describes the steric properties of a molecule by the distance between sampling points 2 Å outside the average van der Waals surface of the molecules, and the closest atom and hydrogen bond acceptor properties by the distance to the complementary ligand atom.

B. CALCULATIONS BASED ON SPATIAL RELATIONSHIPS BETWEEN THE ATOMS OF THE 3-DIMENSIONAL STRUCTURE

Autocorrelation vectors describe the 2D or 3D distribution of atomic or surface properties of a molecule.[149,198–200] The term *autocorrelation* is used because the elements of the vector are calculated from relationships, usually distances, within the molecule itself. Separate autocorrelation vectors describe each molecular property. Each element of such a vector refers to some interpoint distance calculated from the structure. The distances can be the number of bonds that separate atoms in the hydrogen-suppressed 2D structure, 3D interatomic distances, or distances between a number of randomly generated points on the surface of the molecule, etc. Typically, the 2D distances span from 0 to 14 bonds, and 3D distances are binned in 1.0 Å increments from 1 Å to 15.0 Å. Usually, for any one distance increment and molecular property, the value of the vector element is a sum over all such pairs of the product of the property values of each pair of points. For example, each element in a shape autocorrelation vector would be the count of the number of times that particular interatomic distance is present in the structure. Other autocorrelation functions may be generated from atomic lipophilicities, electronegativities, hydrogen-bonding properties, electrostatic properties, or molar refractivities.

The elements of autocorrelation vectors are used as descriptors in a QSAR analysis. For example, we found that a PLS analysis of the antibacterial activity of 28 erythromycin esters using steric autocorrelation vectors produced the same statistics as a CoMFA analysis (reported in Chapter 10) on the same conformation. Autocorrelation vectors are used in such programs as GRIND,[143] Dragon,[201] AutologP,[202] and Adriana.[203]

WHIM and MS-WHIM descriptors[204] transform properties of the atoms and the coordinates of the centered molecule (WHIM) or points on the surface of a molecule (MS-WHIM) into a set of descriptors for each property. For example; properties can be unity (i.e., occupied or not occupied), positive electrostatic potential, negative electrostatic potential, hydrogen bond acceptor, hydrogen bond donor, and hydrophobicity. The first step is to use the covariance matrix of the location and properties of the atomic or surface points in a principal components analysis (discussed in Chapter 7): The algorithm uses these results to calculate the weight, shape, symmetry, and surface property density of each of the six properties with respect to the principal axes of that property to produce a total of approximately 100 descriptors for each molecule.

The CoMMA method calculates the principal moments of the various fields to use as QSAR descriptors in a PLS analysis.[205] Thus, the three shape descriptors are the three principal moments of inertia calculated with a principal components analysis of the atomic positions of the atoms. Principal components analysis of the partial atomic charges on the atoms yields two electrostatic descriptors: the magnitude of the dipole and quadrupole moments. Another set of descriptors is the magnitude of the difference between the center of mass and the center of dipole along the three principal axes of inertia.

Because this class of methods is based on distances between atoms or surface points, they are neutral with respect to chirality. Hence, a special descriptor may be needed to account for enantiomers. An advantage is that if one does not know the bioactive enantiomer, it need not be proposed, nor does one need to specify it for each molecule: Of course, the molecular modeling to suggest the conformation to be used for the analysis may involve selecting an enantiomer to be used.

IX. LESSONS LEARNED

Table 4.12 summarizes the types of descriptors that describe noncovalent interactions for various types of data sets. Many of the simpler methods can also be used with data sets that are more detailed: for example, it is possible to use a calculated log P as one descriptor in an analysis of data sets that contain 3D structures of aligned molecules.

Log D, the apparent log P at a pH at which the compound is partly ionized, is a complex function of the log P of the neutral form and the pK_a of the molecule. The fraction ionized is essentially zero at pHs one log unit away from the pK_a in the direction of the neutral form. At pHs one log unit in the direction of the ionized form, it is a linear function of pK_a.

Indicator variables describe the presence or absence of a particular molecular feature in a molecule. Substructure descriptors are indicator variables, each of which corresponds to a specific possible substructure in the molecule. They are especially useful for the analysis of large data sets of diverse molecules.

Molecular connectivity indices and related descriptors often combine more than one property into a descriptor. They often form the basis of predictive QSARs.

TABLE 4.12

Overview of Methods for Calculating the Strength of Noncovalent Interactions

Structural Information	Type of Noncovalent Interaction			
	Steric	Hydrophobic	Electrostatic	Hydrogen Bonding
2D structures of diverse molecules	MR, MW autocorrelation vector of occupancy	Calculated log P	Calculated pK_a	Abraham sA and sB values, count of O's and N's, 2D PSA
2D structures of a series with a common core	E_s and Sterimol values, Topomer CoMFA (steric)	π or f values, can separate by position	σ values, empirical partial atomic charges, Topomer CoMFA (electrostatic)	σ values, Abraham α_2^H, β_2^H, sA, and sB values
3D structures of a set of molecules in some consistent conformation	Autocorrelation vectors and MS-WHIM based on occupancy	Autocorrelation vectors and MS-WHIM based on atom hydrophobicity, molecular lipophilicity potential	Autocorrelation vectors and MS-WHIM based on electrostatic potential from empirical or quantum chemical atomic charges	Autocorrelation vectors and MS-WHIM based on hydrogen-bonding potential, polar surface area
3D structures of a set of aligned molecules	Force-field steric energies (CoMFA or GRID) or occupancy of certain regions of space	Force-field hydrogen-bonding fields (CoMFA), water probe (GRID)	Force-field CoMFA or GRID electrostatic energies, electrostatic potential from quantum chemistry	Force-field CoMFA, GRID, or HINT hydrogen bond energies
3D structures of the ligand–macromolecule complex	Force-field steric energy—either total or partitioned by protein residue or atom	Implied in the steric and hydrogen-bonding energy—either total or partitioned by protein residue or atom	Force-field electrostatic energy—either total or partitioned by protein residue or atom	Force-field hydrogen-bonding energy—either total or partitioned by protein residue or atom

APPENDIX 4.1. RULES FOR ENCODING A STRUCTURE INTO SMILES

1. Atoms (except hydrogens, which are usually implied) are represented by their atomic symbol enclosed in brackets. Brackets are not needed for F, Cl, Br, I, C, O, N, S, or P if they are not isotopically labeled and do not bear a formal charge or are indicated with an explicit hydrogen. Uppercase denotes aliphatic C, N, O, or S, whereas lowercase denotes aromatic atoms.

2. The default bond between aliphatic atoms is single and that between aromatic atoms in the same ring is aromatic. Double bonds are indicated by a "=", triple bonds by a "#", and a disconnect as in a salt by a ".".

3. Cyclization is indicated by matching numerals following the atomic symbols of the atoms between which the bond is to be formed.

4. Branches are enclosed in parentheses.

Some examples:

Structure 4.28 Ethanol	CCO
Structure 4.29 Acetic acid	CC(=O)O
Structure 4.30 Acetate	CC(=O)[O-]
Structure 4.31 Benzene	c1ccccc1
Structure 4.32 Pyridine	c1ccccn1

(4.28)   (4.29)   (4.30)

(4.31)   (4.32)

REFERENCES

1. Kim, K. H. In *3D QSAR in Drug Design: Theory Methods and Applications.* Kubinyi, H., Ed. ESCOM: Leiden, 1993, pp. 619–42.

2. Hansch, C.; Leo, A.; Hoekman, D. *Exploring QSAR: Hydrophobic, Electronic, and Steric Constants.* American Chemical Society: Washington, DC, 1995.

3. Weininger, D.; Weininger, A. *J. Chem. Inf. Comput. Sci.* **1988**, *28*, 31–6.

4. Livingstone, D. J. *J. Chem. Inf. Comput. Sci.* **2000**, *40*, 195–209.

5. Platts, J. A.; Butina, D.; Abraham, M. H.; Hersey, A. *J. Chem. Inf. Comput. Sci.* **1999**, *39*, 835–45.

6. Weininger, D.; Weininger, A.; Weininger, J. L. *J. Chem. Inf. Comput. Sci.* **1989**, *29*, 97–101.

7. Weininger, D. *J. Chem. Inf. Comput. Sci.* **1990**, *30*, 237–43.

8. http://www.daylight.com/smiles/index.html; 2007. http://www.daylight.com/smiles/index.html (accessed Nov. 25, 2007). Daylight Chemical Information Systems, Inc.

9. ChemDraw, version Pro 11.0. CambridgeSoft. Cambridge, MA. 2007. http://www.cambridgesoft.com/software (accessed Nov. 1, 2007).

10. Ertl, P.; Jacob, O. *J. Mol. Struct.* **1997**, *419*, 113–20.
11. Kubinyi, H., Ed. *3D QSAR in Drug Design: Theory Methods and Applications*. ESCOM: Leiden, 1993.
12. Kubinyi, H.; Folkers, G.; Martin, Y. C., Ed. *3D QSAR in Drug Design. Vol. 2. Ligand–Protein Interactions and Molecular Similarity*. ESCOM: Leiden, 1998.
13. Kubinyi, H.; Folkers, G.; Martin, Y. C., Ed. *3D QSAR in Drug Design. Vol. 3. Recent Advances*. ESCOM: Leiden, 1998.
14. Hansch, C.; Leo, A. In *Exploring QSAR: Fundamentals and Applications in Chemistry and Biology*. American Chemical Society: Washington, DC, 1995, pp. 1–68.
15. Shorter, J. *Correlation Analysis in Organic Chemistry: An Introduction to Linear Free-Energy Relationships*. Clarendon Press: Oxford, 1973.
16. Taft, R. W.; Lewis, I. C. *J. Am. Chem. Soc.* **1958**, *80*, 2436–43.
17. Swain, C. G.; Lupton Jr., E. C. *J. Am. Chem. Soc.* **1968**, *90*, 4328–37.
18. Topliss, J. G.; Edwards, R. P. *J. Med. Chem.* **1979**, *22*, 1238–44.
19. Charton, M. *Prog. Phys. Org. Chem.* **1971**, *8*, 235–318.
20. Jaffe, H. H.; Jones, H. L. In *Advances in Heterocyclic Chemistry*; Katritzky, A. R., Ed. Academic Press: New York, 1963, Vol. 3, pp. 209–61.
21. Taft, R. W. In *Steric Effects in Organic Chemistry*; Newman, M. S., Ed. Wiley: New York, 1956, pp. 556–675.
22. Hall, J., H. K. *J. Am. Chem. Soc.* **1956**, *78*, 2570–2.
23. Hall, H. K. *J. Am. Chem. Soc.* **1957**, *79*, 5444–7.
24. Hall, J., H. K. *J. Am. Chem. Soc.* **1957**, *79*, 5441–7.
25. Clark, J.; Perrin, D. D. *Quarterly Rev. Chem. Soc.* **1964**, *18*, 295–320.
26. Leffler, J. E.; Grunwald, E. *Rates and Equilibria of Organic Reactions as Treated by Statistical, Thermodynamic, and Extrathermodynamic Methods*. Dover: New York, 1963.
27. Momany, F. A.; McGuire, R. F.; Burgess, A. W.; Scheraga, H. A. *J. Phys. Chem.* **1975**, *79*, 2361–81.
28. Gasteiger, J.; Marsili, M. *Tetrahedron Lett.* **1978**, 3181–4.
29. Gasteiger, J.; Marsili, M. *Tetrahedron* **1980**, *36*, 3219–88.
30. Goodford, P. J. *J. Med. Chem.* **1985**, *28*, 849–57.
31. Seibel, G. L.; Kollman, P. A. In *Comprehensive Medicinal Chemistry*; Ramsden, C. A., Ed. Pergamon: Oxford, 1990. Vol. 4, pp. 125–38.
32. Jorgensen, W. L.; Tiradorives, J. *Perspect. Drug Discovery Des.* **1995**, *3*, 123–38.
33. Halgren, T. A. *J. Comput. Chem.* **1996**, *17*, 520–52.
34. Czodrowski, P.; Dramburg, I.; Sotriffer, C. A.; Klebe, G. *Proteins* **2006**, *65*, 424–37.
35. Zhang, J. H.; Kleinoder, T.; Gasteiger, J. *J. Chem. Inf. Model.* **2006**, *46*, 2256–66.
36. Polanski, J.; Gasteiger, J.; Wagener, M.; Sadowski, J. *Quant. Struct.-Act. Relat.* **1998**, *17*, 27–36.
37. Cramer, C. J. Essentials of Computational Chemistry; Theories and Methods. Wiley: New York, 2002, pp. 278–89.
38. Kim, K. H.; Martin, Y. C. *J. Org. Chem.* **1991**, *56*, 2723–9.
39. Alex, A. A. In *Comprehensive Medicinal Chemistry II*; Mason, J. S., Ed. Elsevier: Oxford, 2007, Vol. 4, pp. 379–419.
40. Mulliken, R. S. *J. Chem. Phys.* **1955**, *23*, 1833–40.
41. Dewar, M. J. S.; Zoebish, E. G.; Healy, E. F.; Stewart, J. J. P. *J. Am. Chem. Soc.* **1985**, *107*, 3902–9.
42. Boys, S. F. *Proc. R. Soc. London Ser. A* **1950**, *200*, 542.
43. Weiner, P. K.; Langridge, R.; Blaney, J. M.; Schaefer, R.; Kollman, P. A. *Proc. Natl. Acad. Sci. U. S. A.* **1982**, *79*, 3754–8.
44. Kim, K. H.; Martin, Y. C. *J. Med. Chem.* **1991**, *34*, 2056–60.
45. Kroemer, R. T.; Hecht, P.; Liedl, K. R. *J. Comput. Chem.* **1996**, *17*, 1296–308.

46. Jung, D.; Floyd, J.; Gund, T. M. *J. Comput. Chem.* **2004**, *25*, 1385–99.
47. Jakalian, A.; Bush, B. L.; Jack, D. B.; Bayly, C. I. *J. Comput. Chem.* **2000**, *21*, 132–46.
48. Jakalian, A.; Jack, D. B.; Bayly, C. I. *J. Comput. Chem.* **2002**, *23*, 1623–41.
49. Sillen, L. G.; Martell, A. E. *Stability Constants of Metal–Ion Complexes. Special Publication No. 17. The Chemical Society.* Burlington House: London, 1964.
50. Sillen, L. G.; Martell, A. E. *Stability Constants of Metal–Ion Complexes. Special Publication No. 25. The Chemical Society.* Burlington House: London, 1971.
51. Perrin, D. D. *Dissociation Constants of Organic Bases in Aqueous Solution.* Butterworths: London, 1965.
52. PhysProp Database, Syracuse Research Corporation. Syracuse. 2007. http://www.syrres.com/esc/physdemo.htm (accessed Nov. 1, 2007).
53. Bio-Loom, Biobyte. Claremont CA. 2007. http://biobyte.com/bb/prod/bioloom.html (accessed Nov. 2, 2007).
54. Perrin, D. D. *J. Chem. Soc.* **1965**, 5590–6.
55. Barlin, G. B.; Perrin, D. D. *Quarterly Rev. Chem. Soc.* **1966**, *20*, 75–101.
56. Albert, A. In *Physical Methods in Hetrocyclic Chemistry*; Katritzky, A. R., Ed. Academic Press: New York, 1963, Vol. 3, pp. 1–26.
57. pK_a, CompuDrug. Budapest. http://www.compudrug.com/ (accessed Aug. 2, 2008).
58. ChemAxon Plug-ins, ChemAxon. Budapest. http://www.chemaxon.com/product/calc_pl_land.html.
59. Hilal, S. H.; Karickhoff, S. W.; Carreira, L. A. *Quant. Struct.-Act. Relat.* **1995**, *14*, 348–55.
60. Csizmadia, F.; Tsantilikakoulidou, A.; Panderi, I.; Darvas, F. *J. Pharm. Sci.* **1997**, *86*, 865–71.
61. Tsantilikakoulidou, A.; Panderi, I.; Csizmadia, F.; Darvas, F. *J. Pharm. Sci.* **1997**, *86*, 1173–9.
62. Petrauskas, A. A.; Dolovanov, E. A. *Perspect. Drug Discovery Des.* **2000**, *19*, 99–116.
63. Japertas, P.; Didziapetris, R.; Petrauskas, A. *Quant. Struct.–Act. Relat.* **2002**, *21*, 23–37.
64. Jorgensen, W. L.; Duffy, E. M. *Adv. Drug Delivery Rev.* **2002**, *54*, 355–66.
65. ACD/pK_a DB, version 8.0. Advanced Chemistry Development, Inc. Toronto. 2007. http://www.acdlabs.com/products/phys_chem_lab/ (accessed Nov. 1, 2007).
66. Milletti, F.; Storchi, L.; Sforma, G.; Cruciani, G. *J. Chem. Inf. Model.* **2007**, *47*, 2172–2181.
67. Shelley, J. C.; Cholleti, A.; Frye, L. L.; Greenwood, J. R.; Timlin, M. R.; Uchimaya, M. *J. Comp-Aid. Mol. Des.* **2007**, *21*, 681–91.
68. Lee, A.C.; Crippen, G.M.; *J. Chem. Inf. Model.* **2009**, *23*, 2013–2033.
69. Hansch, C.; Leo, A. *Exploring QSAR: Fundamentals and Applications in Chemistry and Biology.* American Chemical Society: Washington, DC, 1995.
70. Leo, A.; Hoekman, D. *Perspect. Drug Discovery Des.* **2000**, 19–38.
71. Meylan, W. M.; Howard, P. H. *Perspect. Drug Discovery Des.* **2000**, *19*, 67–84.
72. Wang, R.; Gao, Y.; Lai, L. *Perspect. Drug Discovery Des.* **2000**, *19*, 47–66.
73. Molnár, L.; Keseru, G. M.; Papp, Á.; Gulyás, Z.; Darvas, F. *Bioorg. Med. Chem. Lett.* **2004**, *14*, 851–3.
74. Lipinski, C. A.; Lombardo, F.; Dominy, B. W.; Feeney, P. J. *Adv. Drug Delivery Rev.* **1997**, *23*, 3–25.
75. Palm, K.; Stenberg, P.; Luthman, K.; Artursson, P. *Pharm. Res.* **1997**, *14*, 568–71.
76. Ertl, P.; Rohde, B.; Selzer, P. *J. Med. Chem.* **2000**, *43*, 3714–7.
77. Abraham, M. H.; Platts, J. A. *J. Org. Chem.* **2001**, *66*, 3484–91.
78. Absolv, Pharma Algorithms. Toronto. 2008. http://pharma-algorithms.com/absolv.htm (accessed Aug. 2, 2008).

79. Raevsky, O. A. *J. Phys. Org. Chem.* **1997**, *10*, 405–13.
80. HYBOT-PLUS, TimTec. Newark, Delaware. 2008. http://www.timtec.net/software/ hybot-plus.htm (accessed Aug. 2, 2008).
81. Kenny, P. W. *J. Chem. Soc., Perkin Trans. 2* **1994**, 199–202.
82. Boobbyer, D. N.; Goodford, P. J.; McWhinnie, P. M.; Wade, R. C. *J. Med. Chem.* **1989**, *32*, 1083–94.
83. Kim, K. H.; Greco, G.; Novellino, E.; Silipo, C.; Vittoria, A. *J. Comput.-Aided Mol. Des.* **1993**, *7*, 263–80.
84. Wade, R. C.; Clark, K. J.; Goodford, P. J. *J. Med. Chem.* **1993**, *36*, 140–7.
85. Wade, R. C.; Goodford, P. J. *J. Med. Chem.* **1993**, *36*, 148–56.
86. Cornell, W. D.; Cieplak, P.; Bayly, C. I.; Gould, I. R.; Merz, K. M.; Ferguson, D. M.; Spellmeyer, D. C.; Fox, T.; Caldwell, J. W.; Kollman, P. A. *J. Am. Chem. Soc.* **1995**, *117*, 5179–97.
87. Silipo, C.; Vittoria, A. In *Comprehensive Medicinal Chemistry: The Rational Design, Mechanistic Study and Therapeutic Application of Chemical Compounds*; Ramsden, C. A., Ed. Pergamon: Oxford, 1990, Vol. 4, pp. 153–204.
88. Mannhold, R.; Rekker, R. F.; Dross, K.; Bijloo, G.; de Vries, G. *Quant. Struct.-Act. Relat.* **1998**, *17*, 517–36.
89. Kim, K. H.; Martin, Y. C. In *QSAR: Rational Approaches to the Design of Bioactive Compounds*; Silipo, C., Vittoria, A., Eds. Elsevier: Amsterdam, 1991, pp. 151–4.
90. Hancock, C. K.; Meyers, E. A.; Yager, B. J. *J. Am. Chem. Soc.* **1961**, *83*, 4211–3.
91. Charton, M. *J. Am. Chem. Soc.* **1969**, *91*, 615–8.
92. Kutter, E.; Hansch, C. *J. Med. Chem.* **1969**, *12*, 647–52.
93. Babadjamian, A.; Chanon, M.; Gallo, R.; Metzger, J. *J. Am. Chem. Soc.* **1973**, *95*, 3807–8.
94. Verloop, A.; Hoogenstraaten, W.; Tipker, J. In *Drug Design*; Ariens, E. J., Ed. Academic Press: New York, 1976, Vol. 7, pp. 164–207.
95. Cramer III, R. D.; Patterson, D. E.; Bunce, J. D. *J. Am. Chem. Soc.* **1988**, *110*, 5959–67.
96. Cruciani, G.; Watson, K. A. *J. Med. Chem.* **1994**, *37*, 2589–601.
97. Tropsha, A.; Cho, S. J. In *3D QSAR in Drug Design*: Vol. 3, Kubinyi, H., Folkers, G., Martin, Y. C., Eds. Kluwer: Dordrecht, 1997, pp. 57–69.
98. Tropsha, A.; Cho, S. J. *Perspect. Drug Discovery Des.* **1998**, *12*, 57–69.
99. Sprague, P. W. *Perspect. Drug Discovery Des.* **1995**, *3*, 1–20.
100. Dixon, S. L.; Smondyrev, A. M.; Knoll, E. H.; Rao, S. N.; Shaw, D. E.; Friesner, R. A. *J. Comp-Aid. Mol. Des.* **2006**, *20*, 647–71.
101. Hansch, C.; Leo, A. In *Exploring QSAR: Fundamentals and Applications in Chemistry and Biology*. American Chemical Society: Washington, DC, 1995, pp. 97–168.
102. Tanford, C. *The Hydrophobic Effect: Formation of Micelles and Biological Membranes*. Wiley-Interscience: New York, 1973.
103. Davis, S. S.; Higuchi, T.; Rytting, J. H. In *Advances in Pharmaceutical Sciences*; Bean, H. S., Beckett, A. H., Carless, J. E., Eds. Academic Press: London, 1974, pp. 73–261.
104. Smith, R. N.; Hansch, C.; Ames, M. *J. Pharm. Sci.* **1975**, *64*, 599–606.
105. Seiler, P. *Eur. J. Med. Chem.* **1974**, *9*, 473–9.
106. Taft, R. W.; Abraham, M. H.; Famini, G. R.; Doherty, R. M.; Abboud, J. L.; Kamlet, M. J. *J. Pharm. Sci.* **1985**, *74*, 807–14.
107. Hansch, C.; Leo, A. In *Exploring QSAR: Fundamentals and Applications in Chemistry and Biology*. American Chemical Society: Washington, DC, 1995, pp. 100.
108. Collander, R. *Acta Chem. Scand.* **1951**, *5*, 774–80.
109. Leo, A.; Hansch, C. *J. Org. Chem.* **1971**, *36*, 1539–44.
110. Hansch, C.; Leo, A. In *Exploring QSAR: Fundamentals and Applications in Chemistry and Biology*. American Chemical Society: Washington, DC, 1995, pp. 100–3.

111. Pagliara, A.; Caron, G.; Lisa, G.; Fan, W. Z.; Gaillard, P.; Carrupt, P. A.; Testa, B.; Abraham, M. H. *J. Chem. Soc., Perkin Trans. 2* **1997**, 2639–43.
112. Hansch, C.; Leo, A. In *Exploring QSAR: Fundamentals and Applications in Chemistry and Biology.* American Chemical Society: Washington, DC, 1995, pp. 118–21.
113. Avdeef, A.; Barrett, D. A.; Shaw, P. N.; Knaggs, R. D.; Davis, S. S. *J. Med. Chem.* **1996**, *39*, 4377–81.
114. Demotes-Mainard, F.; Jarry, C.; Thomas, J.; Dallet, P. *J. Liq. Chromatogr.* **1991**, *14*, 795–805.
115. Ford, H. J.; Merski, C. L.; Kelley, J. A. *J. Liq. Chromatogr.* **1991**, *14*, 3365–86.
116. Ong, S. W.; Liu, H. L.; Pidgeon, C. *J. Chromatogr.* **1996**, *728*, 113–28.
117. Valko, K.; Bevan, C.; Reynolds, D. *Anal. Chem.* **1997**, *69*, 2022–9.
118. Caldwell, G. W.; Masucci, J. A.; Evangelisto, M.; White, R. *J. Chromatogr.* **1998**, *800*, 161–9.
119. Kansy, M.; Senner, F.; Gubernator, K. *J. Med. Chem.* **1998**, *41*, 1007–10.
120. Valko, K.; Du, C. M.; Bevan, C.; Reynolds, D. P.; Abraham, M. H. *Curr. Med. Chem.* **2001**, *8*, 1137–46.
121. Tomlinson, E. *J. Chromatogr.* **1975**, *113*, 1–45.
122. Dross, K.; Rekker, R. F.; de Vries, G.; Mannhold, R. *Quant. Struct.-Act. Relat.* **1998**, *17*, 549–57.
123. LogKow/KowWin, Syracuse Research Corporation. Syracuse. 2007. http://www.syrres.com/esc/est_kowdemo.htm (accessed Nov. 1, 2007).
124. van de Waterbeemd, H.; Mannhold, R. *Quant. Struct.–Act. Relat.* **1996**, *15*, 410–2.
125. Duban, M. E.; Bures, M. G.; DeLazzer, J.; Martin, Y. C. In *Pharmacokinetic Optimization in Drug Research*; Testa, B., van de Waterbeemd, H., Folkers, G., Guy, R., Ed. Wiley-VCH: Zurich, 2001, pp. 485–97.
126. Mannhold, R.; van de Waterbeemd, H. *J. Comput.-Aided Mol. Des.* **2001**, *15*, 337–54.
127. Fujita, T.; Iwasa, J.; Hansch, C. *J. Am. Chem. Soc.* **1964**, *86*, 5175–80.
128. Norrington, F.; Hyde, R. M.; Williams, S. G.; Wootton, R. *J. Med. Chem.* **1975**, *18*, 604–7.
129. Nys, G. G.; Rekker, R. F. *Chimica Therapeutica* **1973**, *9*, 521–35.
130. Nys, G. G.; Rekker, R. F. *Chimica Therapeutica* **1974**, *10*, 361–75.
131. Mannhold, R.; Rekker, R. *Perspect. Drug Discovery Des.* **2000**, *18*, 1–18.
132. Hansch, C.; Leo, A. In *Exploring QSAR: Fundamentals and Applications in Chemistry and Biology.* American Chemical Society: Washington, DC, 1995, pp. 125–68.
133. CLOGP, version 4.3. Biobyte. Claremont, CA. 2002. http://www.biobyte.com/ (accessed Nov. 1, 2007).
134. Leo, A. J. *Chem. Rev.* **1993**, *93*, 1281–306.
135. Moriguchi, I.; Hirono, S.; Liu, Q.; Nakagome, I.; Matsushita, Y. *Chem. Pharm. Bull.* **1992**, *40*, 127–30.
136. TRIPOS, Inc.: 1699 S. Hanley Road, St Louis, MO 63944.
137. Ghose, A. K.; Pritchett, A.; Crippen, G. M. *J. Comput. Chem.* **1988**, *9*, 80–90.
138. Heiden, W.; Moeckel, G.; Brickmann, J. *J. Comput.-Aided Mol. Des.* **1993**, *7*, 503–14.
139. Wang, R. X.; Fu, Y.; Lai, L. H. *J. Chem. Inf. Comput. Sci.* **1997**, *37*, 615–21.
140. Kim, K. H. *Med. Chem. Res.* **1991**, *1*, 259–64.
141. von Itzstein, M.; Dyason, J. C.; Oliver, S. W.; White, H. F.; Wu, W. Y.; Kok, G. B.; Pegg, M. S. *J. Med. Chem.* **1996**, *39*, 388–91.
142. Cruciani, G.; Pastor, M.; Guba, W. *Eur. J. Pharm. Sci.* **2000**, *11*, S29–S39.
143. Pastor, M.; Cruciani, G.; McLay, I.; Pickett, S.; Clementi, S. *J. Med. Chem.* **2000**, *43*, 3233–43.
144. Kellogg, G. E.; Semus, S. F.; Abraham, D. J. *J. Comput.-Aided Mol. Des.* **1991**, *5*, 545–52.
145. Spyrakis, F.; Amadasi, A.; Fornabaio, M.; Abraham, D. J.; Mozzarelli, A.; Kellogg, G. E.; Cozzini, P. *Eur. J. Med. Chem.* **2007**, *42*, 921–33.

146. Fornabaio, M.; Cozzini, P.; Mozzarelli, A.; Abraham, D. J.; Kellogg, G. E. *J. Med. Chem.* **2003**, *46*, 4487–500.

147. Gaillard, P.; Carrupt, P.-A.; Testa, B. *J. Mol. Graphics* **1994**, *12*, 73.

148. Gaillard, P.; Carrupt, P.-A.; Testa, B.; Boudon, A. *J. Comput.-Aided Mol. Des.* **1994**, *8*, 83–96.

149. Moreau, G.; Broto, P. *Nouv. Chim.* **1984**, *11*, 127–40.

150. Barreca, M. L.; Carotti, A.; Carrieri, A.; Chimirri, A.; Monforte, A. M.; Calace, M. P.; Rao, A. *Bioorg. Med. Chem.* **1999**, *7*, 2283–92.

151. Ooms, F.; Wouters, J.; Collin, S.; Durant, F.; Jegham, S.; George, P. *Bioorg. Med. Chem. Lett.* **1998**, *8*, 1425–30.

152. ADME Boxes, Pharma Algorithms. Toronto. http://pharma-algorithms.com/adme_boxes.htm.

153. Meylan, W. M.; Howard, P. H. *J. Pharm. Sci.* **1995**, *84*, 83–92.

154. Ghose, A. K.; Crippen, G. M. *J. Chem. Inf. Comput. Sci.* **1987**, *27*, 21–35.

155. Ghose, A. K.; Viswanadhan, V. N.; Wendoloski, J. J. *J. Phys. Chem.* **1998**, *102*, 3762–72.

156. Wildman, S. A.; Crippen, G. M. *J. Chem. Inf. Comput. Sci.* **1999**, *39*, 868–73.

157. SAVOL3, College of Pharmacy, The University of Texas at Austin, Austin TX. Pearlman@vax.phr.utexas.edu. Austin TX. 1995.

158. Pipeline Pilot, version 5.1.0.100. SciTegic. San Diego, CA. 2005.

159. ACD/ChemSketch 11.0 Freeware, Advanced Chemistry Development. Toronto. 2008. http://www.acdlabs.com/download/chemsk.html (accessed Aug. 2, 2008).

160. Draper, N. R.; Smith, H. *Applied Regression Analysis*; Wiley: New York, 1966.

161. *SSKeys_whitepaper;* 2008. http://www.mdl.com/solutions/white_papers/SSKeys_white-paper.jsp (accessed May 3, 2008).

162. Roberts, G.; Myatt, G. J.; Johnson, W. P.; Cross, K. P.; Blower, P. E. *J. Chem. Inf. Comput. Sci.* **2000**, *40*, 1302–14.

163. *Fingerprints—Screening and Similarity;* 2008. http://www.daylight.com/dayhtml/doc/theory/theory.finger.html (accessed Aug. 3, 2008). Daylight Chemical Information Systems, Inc.

164. *SciTegic Chemistry Components* 2008. http://accelrys.com/products/datasheets/chemistry-component-collection.pdf (accessed May 4, 2008). Accelrys.

165. Hall, L. H.; Kier, L. B. In *Reviews of Computational Chemistry*; Lipkowitz, K. B., Boyd, D. B., Eds. VCH: New York, 1991, Vol. 2, pp. 367–422.

166. Kellogg, G. E.; Kier, L. B.; Gaillard, P.; Hall, L. H. *J. Comp-Aid. Mol. Des.* **1996**, *10*, 513–20.

167. Kier, L. B.; Hall, L. H. *J. Pharm. Sci.* **1981**, *70*, 583–589.

168. Kier, L. B. *Quant. Struct.-Act. Relat.* **1985**, *4*, 109–116.

169. Hall, L. H.; Mohney, B. K.; Kier, L. B. *J. Chem. Inf. Comput. Sci.* **1991**, *31*, 76–82.

170. Molconn-Z, version 4.00. Edusoft. Richmond VA. 2007. http://www.edusoft-lc.com/molconn/ (accessed Nov. 1, 2007).

171. Wang, W.; Donini, O.; Reyes, C. M.; Kollman, P. A. *Annu. Rev. Biophys. Biomol. Struct.* **2001**, *30*, 211–43.

172. Böhm, H. J. *J. Comput.–Aided Mol. Des.* **1992**, *6*, 61–78.

173. Holloway, M. K.; Wai, J. M.; Halgren, T. A.; Fitzgerald, P. M. D.; Vacca, J. P.; Dorsey, B. D.; Levin, R. B.; Thompson, W. J.; Chen, L. J.; deSolms, S. J.; Gaffin, N.; Ghosh, A. K.; Giuliani, E. A.; Graham, S. L.; Guare, J. P.; Hungate, R. W.; Lyle, T. A.; Sanders, W. M.; Tucker, T. J.; Wiggins, M.; Wiscount, C. M.; Woltersdorf, O. W.; Young, S. D.; Darke, P. L.; Zugay, J. A. *J. Med. Chem.* **1995**, *38*, 305–17.

174. Ortiz, A. R.; Pisabarro, M. T.; Gago, F.; Wade, R. C. *J. Med. Chem.* **1995**, *38*, 2681–91.

175. Head, R. D.; Smythe, M. L.; Oprea, T. I.; Waller, C. L.; Green, S. M.; Marshall, G. R. *J. Am. Chem. Soc.* **1996**, *118*, 3959–69.

176. Wang, T.; Wade, R. C. *J. Med. Chem.* **2001**, *44*, 961–71.
177. Gohlke, H.; Klebe, G. *J. Med. Chem.* **2002**, *45*, 4153–70.
178. Jansen, J. M.; Martin, E. J. *Curr. Opin. Chem. Biol.* **2004**, *8*, 359–64.
179. Holloway, M. K. In *3D QSAR in Drug Design*: Vol. 2; Kubinyi, H., Folkers, G., Martin, Y. C., Eds. Kluwer: Dordrecht, 1997, pp. 63–84.
180. Li, Z.; Lazaridis, T. *Phys. Chem. Chem. Phys.* **2007**, *9*, 573–81.
181. Pastor, M.; Cruciani, G.; Watson, K. A. *J. Med. Chem.* **1997**, *40*, 4089–102.
182. Wang, T.; Wade, R. C. *J. Med. Chem.* **2002**, *45*, 4828–37.
183. Lu, Y. P.; Yang, C. Y.; Wang, S. M. *J. Am. Chem. Soc.* **2006**, *128*, 11830–9.
184. Androulakis, I. P.; Nayak, N. N.; Ierapetritou, M. G.; Monos, D. S.; Floudas, C. A. *Proteins: Struct., Funct., Genet.* **1997**, *29*, 87–102.
185. Ferrara, P.; Gohlke, H.; Price, D. J.; Klebe, G.; Brooks, C. L. *J. Med. Chem.* **2004**, *47*, 3032–47.
186. Ortiz, A. R.; Pisabarro, M. T.; Gago, F.; Wade, R. C. *J. Med. Chem.* **1997**, *40*, 1136–1148.
187. Klebe, G.; Mietzner, T.; Weber, F. *J. Comput.-Aided Mol. Des.* **1999**, *13*, 35–49.
188. Mitchell, J. B. O.; Laskowski, R. A.; Alex, A.; Forster, M. J.; Thornton, J. M. *J. Comput. Chem.* **1999**, *20*, 1177–85.
189. Mitchell, J. B. O.; Laskowski, R. A.; Alex, A.; Thornton, J. M. *J. Comput. Chem.* **1999**, *20*, 1165–76.
190. Muegge, I.; Martin, Y. C. *J. Med. Chem.* **1999**, *42*, 791–804.
191. Muegge, I.; Martin, Y. C.; Hajduk, P. J.; Fesik, S. W. *J. Med. Chem.* **1999**, *42*, 2498–503.
192. Warren, G., L.; Andrews, C. W.; Capelli, A. M.; Clarke, B.; La Londe, J.; Lambert, M. H.; Lindvall, M.; Nevins, N.; Semus, S. F.; Senger, S.; Tedesco, G.; Wall, I. D.; Woolven, J. M.; Peishoff, C. E.; Head, M. S. *J. Med. Chem.* **2006**, *49*, 5912–31.
193. Silber, K.; Heidler, P.; Kurz, T.; Klebe, G. *J. Med. Chem.* **2005**, *48*, 3547–63.
194. Hillebrecht, A.; Supuran, C. T.; Klebe, G. *ChemMedChem* **2006**, *1*, 839–53.
195. Klebe, G.; Abraham, U.; Mietzner, T. *J. Med. Chem.* **1994**, *37*, 4130–46.
196. Doweyko, A. M. *J. Med. Chem.* **1988**, *31*, 1396–406.
197. Jain, A. N.; Dietterich, T. G.; Lathrop, R. H.; Chapman, D.; Critchlow, R. E.; Bauer, B. E.; Webster, T. A.; LozanoPerez, T. *J. Comput.-Aided Mol. Des.* **1994**, *8*, 635–52.
198. Broto, P.; Devillers, J. In *Practical Applications of Quantitative Structure–Activity Relationships (QSAR) in Environmental Chemistry and Toxicology*; Karcher, W., Devillers, J., Eds. Kluwer: Dordrecht, 1990, pp. 105–27.
199. Clementi, S.; Cruciani, G.; Riganelli, D.; Valigi, R.; Costantino, G.; Baroni, M.; Wold, S. *Pharm. Pharmacol. Lett.* **1993**, *3*, 5–8.
200. Wagener, M.; Sadowski, J.; Gasteiger, J. *J. Am. Chem. Soc.* **1995**, *117*, 7769–75.
201. Todeschini, R.; Consonni, V. *Handbook of Molecular Descriptors*, 1st ed. Wiley-VCH: Weinheim, 2000, Vol. 11.
202. Devillers, J.; Domine, D.; Guillon, C.; Karcher, W. *J. Pharm. Sci.* **1998**, *87*, 1086–90.
203. Adriana, Molecular Networks. Erlangen. 2007. http://www.mol-net.com (accessed Oct. 31, 2007).
204. Todeschini, R.; Gramatica, P. In *3D QSAR in Drug Design*; Kubinyi, H., Folkers, G., Martin, Y. C., Eds. Kluwers, Dordrecht, 1998, Vol. 2, pp. 355–80.
205. Silverman, B. D.; Platt, D. E. *J. Med. Chem.* **1996**, *39*, 2129–40.

5 Biological Data

The precision of predictions from a quantitative structure–activity relationship (QSAR) is correlated with the precision with which its associated biological potency was measured. This chapter will discuss the characteristics of an appropriate biological test and how one converts ordinary biological data into a measure of potency for use in quantitative modeling. Potency will be loosely defined as the reciprocal of the dose or concentration required to produce some predetermined biological response, expressed as $\log (1/C)$, $-\log (C)$, or pC. Thus, the lower the dose required, the higher the potency. In the discussion that follows, dose is used as a generic term for the amount of compound added to a system. It is literally the dose in animal (in vivo) studies, but it also refers to the concentration in test tube (in vitro) experiments.

The relevance of predictions from a QSAR to, for example, cure of a disease depends on the relevance of the biological test used for the QSAR. While most of the responsibility for the design of appropriate tests rests with the biologist, the modeler must not lose sight of the ultimate relevance of the test on which the calculations are based. Careful attention to this point may suggest, for example, that it is equally important to do a rough calculation using in vivo potency and toxicity as it is to develop a regression equation for in vitro potency.

This chapter is a minimal introduction to biological testing. There are other books that may be useful.[1–4]

I. CONSEQUENCES OF LIGAND–BIOMOLECULE INTERACTION

Chapter 2 contains a discussion of the types of interactions that may occur between a ligand and its target biomolecule. How is this ligand–biomolecule interaction transformed into a biological response? The answer to such a question is the realm of biochemical pharmacology. However, a discussion of some common consequences of the ligand–biomolecule interaction may be helpful. Compounds tested in even the apparently simplest in vitro test can produce their effects by more than one means.

- Mimicking the effect of a normal body constituent on its normal target: Such an effect might be prolonged compared to the normal constituent, or it might occur only in certain tissues: Compounds that act this way are agonists.

- Competitively blocking a normal body constituent from its normal target to inhibit the biological consequences of the natural constituent: Compounds that act in this way are inhibitors or antagonists.
- Releasing stored normal body constituents: The initial burst or the subsequent depletion of such a constituent may be responsible for the observed biological effect.
- Changing the conformation of the biological target macromolecule: If this macromolecule is an enzyme, such a change may result in altered substrate specificity or catalytic activity, or both.
- Depolarizing or disrupting a membrane.
- Forming a covalent bond between the compound and the macromolecule: Inhibition of an enzyme or conformational changes may be observed to occur only after the covalent bond has formed.
- Interacting with an enzyme as a pseudosubstrate: The ligand's disruptive effect may be evidenced only after it has been chemically transformed by a normal metabolic sequence.
- Reacting with a normal small molecule or metal ion rather than a macromolecule.
- Inhibiting the enzymes that metabolize natural substances: The result could be an increased effect of the usual amount of such substances.

It must be recognized that a living system is a finely tuned system that contains many feedback loops and communication networks that keep it in check. As a consequence, it can be extremely difficult to tease out the exact mechanism of action of a compound discovered by its biological effect on a whole cell or whole animal. Conversely, it is difficult to predict the overall biological consequences of some isolated biochemical event.

II. SELECTION OF DATA FOR ANALYSIS: CHARACTERISTICS OF AN IDEAL BIOLOGICAL TEST

A. BASED ON A DOSE–RESPONSE CURVE

In general, it is most satisfactory to base structure–activity models on data generated from a test in which a continuous response has been measured at several concentrations. If one considers the response at a single concentration, several compounds might produce the maximum response, but these compounds may differ widely in potency; the same is true of compounds that are inactive at the test concentration but might show varying potency at a higher concentration. A hypothetical example is shown in the curves in Figure 5.1. It can be seen that, at a dose of 1.0, Compounds A and C show a greater response than B, but at a dose of 10, Compounds B and C show a greater response than A. Neither measure reflects the fact that A reaches its maximum response at a lower dose than does B or C.

Dose–response tests allow one to distinguish between the separate effects of (1) affinity of the compound for the biological target, (2) the maximum biological response that can be produced by the compound. For example, consider the data in

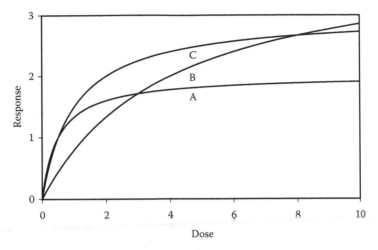

FIGURE 5.1 Hypothetical dose–response curves for three different compounds. Curve A is for a compound with a maximum response of 2.0 and a half-maximum response reached at a concentration of 0.5; curve B is for a compound with a maximum response of 4.0 and a half-maximum response reached at a concentration of 4.0; and curve C is for a compound with a maximum response of 3.0 and a half-maximum response reached at a concentration of 1.0.

Table 5.1.[5] From the left side of the table, it can be seen that, at 1 mM concentration, tyramine (Structure 5.1) is oxidized twice as fast as serotonin (Structure 5.2) and five times faster than octopamine (Structure 5.3). Is tyramine the "best" substrate? The results of a study of the concentration dependence of enzymatic oxidation with the three substrates are shown on the right side of the table. Curves generated from the results are shown in Figure 5.2. They reveal that differences in the rate of reaction is due to differences in both the maximal velocity, V_{max}, and the affinity as measured as the concentration at which half the V_{max} is attained, the K_m. K_m

TABLE 5.1
The Difference between the Results of a Fixed-Concentration and a Dose–Response Study[5]

Substrate	Constant Concentration (1.0 mM Substrate)	Variable Concentration of Substrate	
	mµmol NH$_3$/hr.[a]	K_m (mM)[b]	V_{max} (mµmol NH$_3$/hr.)
Tyramine	164	0.20	197
Serotonin	71	3.84	342
Octopamine	30	2.53	106

[a] Rate of NH$_3$ production by substrates of human liver mitochondrial monoamine oxidase studied at pH 7.40 and 37°C.

[b] K_m is the concentration at which the velocity is half the maximum.

$$HO-\langle\ \rangle-CH_2CH_2NH_2$$

(5.1)

$$HO-[indole]-CH_2CH_2NH_2$$

(5.2)

$$HO-\langle\ \rangle-\overset{OH}{\underset{|}{C}}HCH_2NH_2$$

(5.3)

is 20-fold lower for tyramine than serotonin, whereas the V_{max} for tyramine is approximately one-half that for serotonin. Thus, the complete concentration-rate curves tell us that, depending on the concentration used at any fixed concentration, tyramine may be oxidized at a slower rate than, equal to, or faster than serotonin. Octopamine is always oxidized at a slower rate than tyramine. If one wishes to make sensible interpretations from QSAR models, then it is clearly more relevant to consider the effects of changes in physical properties on affinity and maximal activity separately.

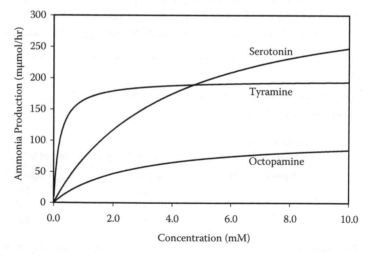

FIGURE 5.2 Plot of the rate of oxidation by monoamine oxidase as a function of concentration of tyramine, serotonin, and octopamine. The curves were calculated from the constants given by McEwen et al.[5] and Equation 5.12.

A complete dose–response curve will also reveal cases of reversal of activity at high concentrations. This is most often seen with central nervous system-type drugs that might, for example, be sedatives at low doses and stimulants at high doses.

Sometimes the actual measurement of biological activity of a compound is the prevention of death in a disease model of fatal infection, malaria, or cancer. In such a test, increased survival at any dose is a hint of activity. Obviously, the animals may die either from the infection or from the toxic effects of the compound. A parallel toxicity test in "healthy" animals that have been dosed with the test compound alone may be useful to sort out the separate effects.

B. BASED ON THE TIME COURSE OF RESPONSE OF IN VIVO STUDIES

In the case of whole animal tests, sometimes a useful measure of relative potency is based on the area under the time–response curve. This is especially useful if the compound is administered orally and the time of peak response varies from analog to analog (or animal to animal with the same compound), or if the compounds differ in duration of action. Thus, each analog is administered at several doses, and the response is measured at several time points after administration of the compound. The area under each time-response curve is calculated, and this area is plotted versus the logarithm of the dose. The relative potency of the various analogs is the reciprocal of the dose required to produce a given area.

The area under the curve is sometimes already integrated in the reported biological results. An example would be diuretic assays in which the total volume of urine excreted in a particular time interval is measured. Otherwise, the area is calculated by summing the areas of the series of trapezoids that make up the time-response curve. This is done by the following equation:

$$\text{Area} = \frac{\sum_{1}^{j} [R_i + R_{i+1}] \times [t_{i+1} - t_i]}{2} \tag{5.1}$$

in which R_i is the response at the ith measurement, and t_i is the time of the ith measurement. The area is then used as the measure of response in the dose–response calculations described in Section III.

The advantage of using the time-response curve is that this is a better measure of the total effect of a compound than the response at any given point in time. Consider the curves in Figure 5.3.[6] If any one time point were used in the calculation, Compound 1 could be more, equal, or less potent than Compound 3. However, the total effects of these compounds are equal.

The disadvantage of using the time-response curve is that it is influenced by factors other than the intrinsic potency of the compound. For example, the absorption, distribution, metabolism, and excretion of the compound influences the maximum and time course of the biological action, and can make it difficult to separate the effects of variation of structural properties on affinity for the target from the effects on pharmacokinetic properties.

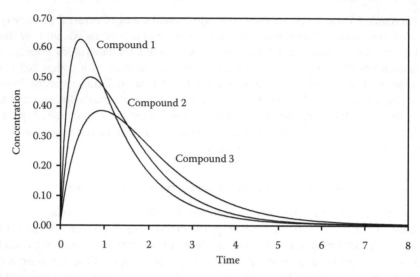

FIGURE 5.3   Blood level curves of identical areas but different shapes. The levels were cal-culated for a one-compartment open model[6] ($C_t = \dfrac{(e^{-Kt} - e^{-k_a t})}{k_a - K} k_a$) with the following relative rate constants: elimination (K) = 1; absorption (k_a) = 1.1, 2, and 4 for Compounds 1, 2, and 3, respectively.

C.   RELEVANT PROPERTIES CONSIDERED

Figure 5.3 again suggests the question of relevancy. If the goal of synthesizing new analogs is to provide a longer duration of action, one could calculate the effect of physical properties on the half-life of biological response as well as the total effect. Comparison of the two equations might suggest analogs with high total effect and the desired half-life.

D.   PRECISION REQUIRED VERSUS POTENCY RANGE

An important consideration for quantitative structure–activity analysis is the pre-cision that is required of the biological assay. The details of the calculations that answer this question are discussed in Chapter 7. However, at this point it can be said that one's chances of finding a statistically significant regression equation increase when one increases the range of potency values within the series, as well as when one increases the precision of individual potency values. Generally speaking, if a dose–response curve has once been run on a compound, and if the potency is estab-lished within ~20%, then additional biological testing need not be done on this com-pound. Instead, the time is better spent establishing the potency of relatively less potent analogs. Both from a statistical and conceptual model-building point of view, it is important to establish the potency of "inactive" compounds so that they may be explicitly included in the structure–activity analysis and their properties included in any forecasts of potency of additional analogs.

E. DEFINITIONS OF SOME PHARMACOLOGICAL TERMS

An *agonist* is a substance that, when administered, results in an observable biological response. The classic agonists are physiological substances such as histamine (Structure 5.4), acetylcholine (Structure 5.5), or methionine encephalin (Structure 5.6). Synthetic agonists are also known. An example is muscarine (Structure 5.7), the administration of which results in several of the biological effects of acetylcholine.

An *antagonist* is a substance the administration antagonizes or blocks the effect of an agonist. Antagonists are usually assumed to compete with the agonist for the binding site on the biomolecule. For example, atropine (Structure 5.8) is an antagonist of acetylcholine (Structure 5.5). Both are suggested to bind to the same biomolecule because both contain an ester group and, at biological pHs, both contain a charged amine.

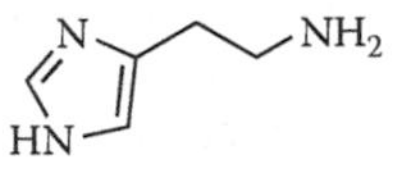

(5.4)

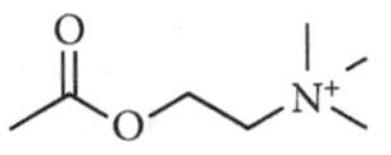

(5.5)

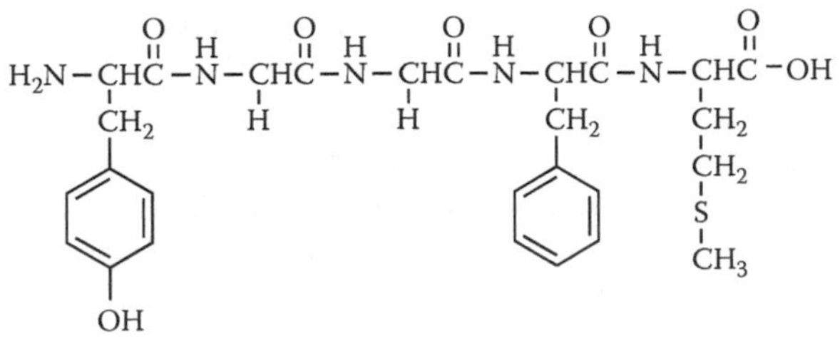

(5.6)

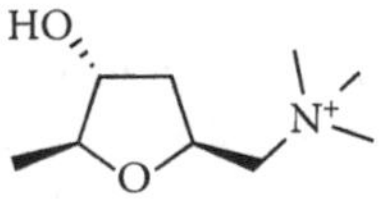

(5.7)

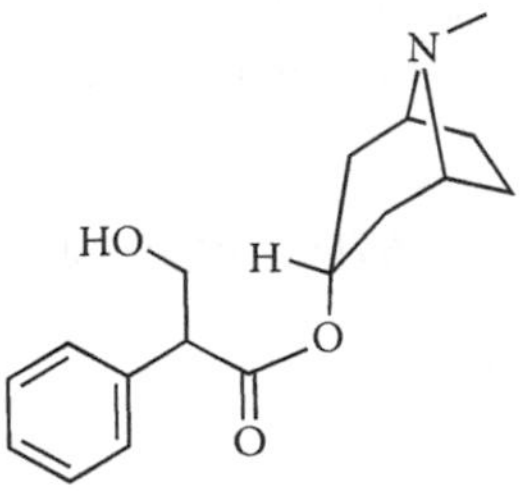

(5.8)

A *partial agonist* is one that, no matter how large a dose is administered, does not produce the maximum biological response exhibited by other known agonists of the system. Such compounds at appropriate doses may antagonize the response of "full" agonists.

The *intrinsic activity* (IA) of an agonist is defined in relative terms as

$$IA_i = \frac{E_{m,i}}{E_{m,X}} \qquad (5.2)$$

in which $E_{m,i}$ is the maximum biological response observed at any dose of Compound i, and $E_{m,X}$ is the maximum biological response observed for the maximally active Compound X in the system under study. The activity of X must be specified in order for the definition to be complete. Thus, the usual maximum intrinsic activity of any compound is 1.0. However, it is sometimes observed that synthetic compounds produce a larger maximum effect than the natural agonist—such compounds are labeled superagonists. The intrinsic activity of partial agonists is less than 1.0.

The *affinity* of an agonist or antagonist for a biological target is a variously defined term that describes the relative dose at which a compound interacts with the target. It takes a larger dose of a compound of low affinity to produce a given effect than to produce the same effect with a compound of high affinity.

ED_{50} (Effective Dose 50%) is the dose of compound required to produce 50% of the maximal effect. There is some fuzziness of definition here because the maximal effect used for comparison may be either that for the compound in question, $E_{m,i}$, or the maximal effect for the best compound in the biological test, $E_{m,X}$. Sometimes ED_{50} is used as the dose that produces 50% increase or decrease in response without regard to the maximum possible increase or decrease. If interpretations are to be made in terms of relative binding to the target, then $E_{m,i}$ is the appropriate comparison. If the total effect is the parameter of interest, then $E_{m,X}$ is the one of choice.

pD_2 is defined for an agonist. It is the negative logarithm of the molar dose at which the effect is one-half the maximum. Thus, in cases in which the dose–response curve is linear, pD_2 is a measure of the affinity of the compound for the target biomolecule. In similar fashion, a pD_{10} is the dose of agonist at which the effect is one-tenth the maximum. The same ambiguity of definition applies to pD_2 as to ED_{50}.

pA_2 is defined for an antagonist. It is measured as the negative logarithm of the concentration of antagonist that requires a doubling of the dose of the agonist to compensate for the action of the antagonist. A pA_{10} value is the negative logarithm of the concentration of antagonist for which it is necessary to increase the agonist concentration 10-fold to restore the original effect.

III. CALCULATION OF RELATIVE POTENCY

A. GENERAL CONSIDERATIONS

Because QSARs consider free energies of interaction, and the free energy of a process is related to the logarithm of the rate or equilibrium constant for the process, for QSAR one would translate the observed response into the logarithm of a relative rate or equilibrium constant. In deciding how to calculate relative potency, one should try to keep in mind what sort of rate or equilibrium process the biological test might represent. The relative potency should be stated in terms of an equilibrium or rate constant if at all possible. Obviously, in the area calculations described in Section II.B, this caveat is not obeyed.

A simple example of the advantage of considering the fundamental biological process is the instance in which the biological data are the relative rates: absorption through the gut, for example. The numbers are often reported as percent of the various analogs absorbed. Without thinking, one might try to correlate the logarithm of the percent absorbed with physical properties. This is incorrect. The rates of biological processes are in general first order, that is,

$$\frac{\partial c_i}{\partial t} = k_i c_i \tag{5.3}$$

in which k_i is the rate constant for Compound i, c_i is its concentration at any instant of time, and t is the time. The integral of Equation 5.3 from time equals zero to time equals t is

$$k_i = \frac{\ln(c_{i,o}/c_i)}{t} = \frac{\ln(1/f_i)}{t} \tag{5.4}$$

in which $c_{i,o}$ is the original concentration of Compound i and $f_i = c_i/c_{i,o}$. If the compounds were all evaluated over the same time interval, that is, t is constant, and if all are absorbed by a first-order process, then the relative rate constants for absorption are *not* proportional to the fraction absorbed but rather to the logarithm of the reciprocal of the fraction remaining at time t. In other words, the rate constant is inversely proportional to the logarithm of the percent *not* absorbed rather than directly proportional to the logarithm of the percent that is absorbed.

Hence, one should transform the biological data to that form that most closely parallels the rate or equilibrium constant appropriate to the test.

B. THEORETICAL DESCRIPTION OF AN IDEALIZED DOSE–RESPONSE CURVE

By making certain simple assumptions, Michaelis and Menten mathematically described the relationships between enzyme concentration, substrate concentration, and the rate of an enzymatic reaction. This is discussed in most standard biochemistry texts. In a similar fashion, the relationship between the dose or concentration of a compound and the observed effect has been formulated.[1–3] The assumptions and derivations are so similar that they will be treated in parallel. It should be noted that the treatment is generally satisfactory for enzyme–substrate or ligand–macromolecule interactions, whereas it is only a rough description of the dose–response characteristics of compounds in complex biological systems.

A compound, usually a small molecule, may be thought to produce its effect by combination with a specific constituent, usually a macromolecule. This vaguely defined constituent with which a compound interacts is often termed the receptor for that ligand, whether or not a true receptor as defined biochemically is involved. In a parallel manner, a substrate, usually a small molecule, interacts with a specific enzyme, a macromolecule. In this instance, the effect is the conversion of substrate into product.

Although certain compounds interact irreversibly with target biomolecules, most act in a reversible fashion. Thus, the following derivation assumes that the reactions follow the law of mass action: that ligands act reversibly. In addition, it is assumed that ligand–biomolecule interaction results in a graded response, the magnitude of which is proportional only to the number of biomolecule sites occupied by the ligand.

According to the foregoing assumptions and those listed in Table 5.2, the reaction between a ligand A and target biomolecule R may be represented as follows:

$$A + R \rightleftharpoons RA \tag{5.5}$$

From its definition K_X, the ligand–biomolecule dissociation constant, is equal to

$$K_X = \frac{[A][R]}{[RA]} \tag{5.6}$$

The square brackets indicate molar concentrations. Because the total concentration of biomolecules, $[R]_t$, is equal to the concentration of free biomolecule plus that bound,

$$[R] = [R]_t - [RA] \tag{5.7}$$

Substituting Equation 5.7 into Equation 5.6,

$$K_X = \frac{[A] \times [R]_t - [A] \times [RA]}{[RA]} \tag{5.8}$$

Solving for [RA],

$$[RA] = \frac{[A] \times [R]_t}{K_X + [A]} \tag{5.9}$$

Assumption 3 of Table 5.2 says that the effect E is proportional to the number of target biomolecules occupied by the ligand, [RA]. From this assumption and Equation 5.9,

$$E = k[RA] = \frac{k[A] \times [R]_t}{K_X + [A]} \tag{5.10}$$

But, from Assumption 4 of Table 5.2, the maximum effect, E_m, is defined as

$$E_m = k[R]_t \tag{5.11}$$

Thus,

$$E = \frac{E_m [A]}{K_X + [A]} \quad \text{or} \quad \frac{E}{E_m} = \frac{[A]}{K_X + [A]} \tag{5.12}$$

TABLE 5.2
Comparison of Quantitation of Enzyme–Substrate and Ligand–Biomolecule Relationships

	Enzyme–Substrate	Ligand–Biomolecule
Symbols	S = Substrate	A = ligand
	E = enzyme	R = target biomolecule
	ES = enzyme–substrate complex	RA = ligand–biomolecule complex
	P, Q = products	E = effect of the ligand
	v = initial velocity of the reaction	
Reaction scheme	$S + E \rightleftharpoons ES \longrightarrow P + Q + E$	$A + R \rightleftharpoons RA \longrightarrow E$
Assumptions	(1) Derivation holds only for the period for which $P + Q + E$ do not form S	One ligand molecule combines with one target biomolecule
	(2) $[ES] \ll [S]$	$[R] \ll [A]$
	(3) $v \propto [ES]$	$E \propto [RA]$
	(4) Maximal velocity V_{max} is reached when $[ES] = [E]_t$	Maximum effect E_m is reached when $[RA] = [R]_t$
Measurements	v as a function of $[S]$ at constant $[E]_t$	E as a function of $[A]$ at constant $[R]_t$
Constants	V_{max} (maximal velocity)	E_m (maximum effect)
Characteristic of the substances	K_m (Michaelis constant, the concentration of substrate at half-maximal velocity)	K_X (ligand–biomolecule dissociation constant, the dose required to produce half the maximum effect)
Equation 5.12 (see Section III.B)	$V = \dfrac{V_{max}[S]}{K_m + [S]}$	$E = \dfrac{E_m[A]}{K_X + [A]}$

Equation 5.12 describes the expected relationship between the observed effect and the concentration of a ligand. This is not a linear relationship over the whole concentration range.

C.   TRANSFORMATIONS OF THE DOSE–RESPONSE CURVE

Equation 5.12 is that of a rectangular hyperbola; a plot of the fractional effect, E/E_m, versus $[A]$ rises from zero when $[A] = 0$ to 1.0 when $[A] = \infty$. Figure 5.4 is a plot of the fraction of maximal expected response at various levels of ligand from 0.10-fold to 10-fold the K_X.

Several transformations may be used to convert Equation 5.12 into a linear relationship,[7] which make it easier to evaluate K_X and E_m. Any of these can be fit by linear regression analyses to solve for the parameters of interest. For example, one may take the reciprocal of Equation 5.12:

$$\frac{1}{E} = \frac{K_X + [A]}{E_m[A]} = \frac{K_X}{E_m[A]} + \frac{1}{E_m} \tag{5.13}$$

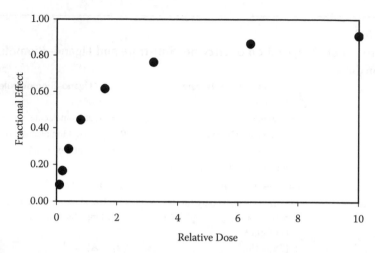

FIGURE 5.4 A plot of the expected response from Equation 5.12 at various levels of the compound. The response is in terms of the fraction of the maximal response, whereas the dose is in multiples of K_X. Doses from 0.10 to 10 times the K_X are shown.

Thus, if one makes a plot of $1/E$ versus $1/[A]$, the intercept of this line on the $1/E$ axis is $1/E_m$, and that on the $1/[A]$ axis is $-1/K_X$. This is shown in Figure 5.5 for the same data as just given. Note that this reciprocal method is not the most reliable way to evaluate E_m and K_X because the most accurately determined responses, those at high concentrations, are all bunched together near the axes. Too much weight is given to the relatively unreliable responses determined at low concentrations.

Another transformation is accomplished by realizing that, if the dose–response curve follows Equation 5.12, then, at doses that give between 20% and 80% of the maximal response, the response is a linear function of the logarithm of the dose. The

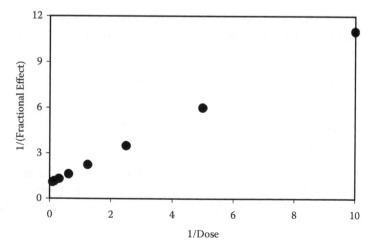

FIGURE 5.5 A plot of the same data as for Figure 5.4, except that the reciprocal of the response and of the dose is plotted.

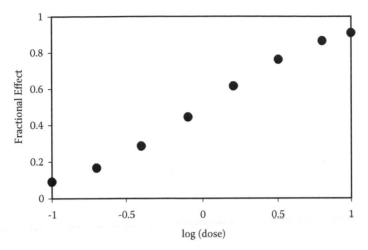

FIGURE 5.6   A plot of the same data as for Figure 5.4, except that the logarithm of the dose is plotted.

sample data is plotted this way in Figure 5.6. From a least-squares regression fit of the log dose *versus* the percent maximum response, it is possible to calculate the dose required to produce a given response. Usually, the response at 50% is used because this dose is proportional to the affinity constant at a constant concentration of the target biomolecule. Thus, one calculates the dose that results in 50% of the maximal response or 50% lethality, the ED_{50} and the LD_{50}, respectively.

A probit plot may also be used for data that fit Equation 5.12. The basis of the probit plot is the assumption that the responsiveness of individual biomolecules follows a normal distribution. The normal distribution has a characteristic center, the mean. Fifty percent of the observations are above, and 50% below the mean. A normal distribution also has a characteristic width parameter, the standard deviation. For example, 84% of the observations lie below the value of the mean plus two standard deviations. Hence, each percentage corresponds to a characteristic number of standard deviations from the mean. For a dose–response curve, the maximum response is given the value of 1.00. The percent of maximum of each of the other responses observed is calculated. The probit is easily calculated in Microsoft Excel with the function NORMINV. A plot of the probit versus the log dose is linear over the range of 5–95% of maximum response. The sample data plotted in this manner is shown in Figure 5.7. It can be seen that this is more linear than in Figure 5.6. Additionally, in contrast to the reciprocal plots, responses at each dose level contribute equally to the slope of the line.

The precision of the estimate of the ED_{50} or LD_{50} values are also provided by any of the regression analyses described thus far. Relatively imprecise estimates may merely reflect variable biological data or they may provide an indication that more than one type or class of binding site is involved.

The log transformation or probit analysis may not be the transformation that best fits the actual data. A recommended procedure is to plot the response versus dose and versus log dose, the probit of response versus log dose, and any other way that seems sensible in light of the biological test. The transformation that best linearizes

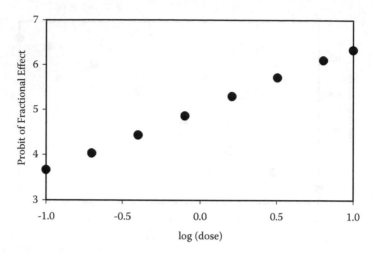

FIGURE 5.7 A plot of the same data as for Figure 5.6, except that the probit of the response and the logarithm of the dose is plotted.

the data would then be used for the actual computer least-squares calculation of the ED_{50} or LD_{50}. Alternately, one may use the computer to calculate the regression lines for the various transformations and compare the results. The important parameters to compare are the width of the 95% confidence interval of the ED_{50} or LD_{50} and the correlation coefficient between the observed and expected responses.

D. RELATIVE POTENCY WITHIN A SERIES: DEFINED ED_{50} OR LD_{50}

For any QSAR method, potency is properly expressed on a molar basis. Thus, an ED_{50} calculated on a milligram per kilogram basis must be converted to some multiple of moles per kilogram. Normally, the relationships are calculated from the relative logarithms of rate or equilibrium constants. Thus, the final potency measure is typically $\log(1/C)$, in which C is the molar concentration required to produce a predefined response. Frequently, the biological potencies are reported relative to a standard. As long as the values are relative molar potencies, the logarithms of these numbers may be used for the calculations.

E. RELATIVE POTENCY WITHIN A SERIES: RESPONSE AT A CONSTANT DOSE

It is usually assumed for a series of analogs that, within specified response limits, the log dose–response curves, such as are seen in Figure 5.6, are parallel lines. This assumption may be expressed in equation form:

$$\log \text{dose}_x = a_x + b \times \text{response}_x \tag{5.14}$$

in which x refers to Compound X, and b is the slope of the log dose versus response curve. It may be assumed that (and verified on a few compounds) the slope b does not vary from compound to compound.

The value of interest for the structure–activity analysis is the logarithm of the relative potency; this is also the negative logarithm of the dose necessary to produce the given response. From Equation 5.14 it may be seen that, at constant response and constant slope b, the logarithm of the dose of Compound X is proportional to the value of a_x, which in turn is proportional to the negative log of the potency. In the special case where all compounds are administered at the same dose, the logarithm of relative potency is proportional to the observed response and *not* proportional to the logarithm of the response.

If the slope of the log dose–response curve is not available, then the proportionality between response and potency will not be known. As a result, in equations that relate such responses to physical properties, the coefficients of the physical properties in these equations have an unknown proportionality to potency. If the QSAR equation identifies optimum values of the physical properties, these are correctly defined.

F. RELATIVE POTENCY WITHIN A SERIES: RESPONSE AT ONE VARIABLE DOSE, AND SLOPE OF DOSE–RESPONSE CURVE KNOWN

In certain instances, biologists will determine the dose–response curve of a few very potent analogs and then report the test results of the other compounds at only a single dose. Because some analogs are inactive at the original test dose, they might have been tested and found active at a second, higher dose. Subject to the limitations discussed in Section III.E, the relative molar potency can be estimated from the data if one assumes that the log dose–response curves of all analogs are parallel. This assumption may be verified with a few analogs. If there is enough variation in potency within the series, a satisfactory structure–activity analysis may be made.

For the calculation, a rearranged form of Equation 5.14 is used:

$$\log \text{potency}_x = -a_x = -\log \text{dose}_x + b \times \text{response}_x \qquad (5.15)$$

The logarithm of the molar dose of the analog and its corresponding response are substituted into the equation for which b has been evaluated from the standard compounds. Only points from 20 to 80% maximal response should be used.

IV. CHOICE OF CLASSIFICATION BOUNDARIES

Chapter 11 will discuss a number of techniques that one can use to develop models that explain the classification of a compound. For example, one might have data that certain compounds are toxic but others are not, or that certain compounds are active but others are not. Classification methods can also be used if one has gradation of potencies such as inactive, weak, active, and potent. The advantage of classification methods is that one does not need a precise measure of the relative potency of various members of the series.

Molecules must be assigned to the various activity classes on the basis of objective criteria, however. There is no problem of classification if the biologist has already provided the data in the form of "active" and "inactive," or +++, ++, +, and 0. The problem occurs when the results are reported as numbers that are known to be imprecise estimates. How are the analogs divided into classes? Where is the line between "active" and "inactive?"

Ideally, the classification criteria assign compounds into natural groupings. These groupings may be discovered from a frequency distribution of response level, that is, a plot of number of examples versus response level. Natural groups and the cutoff points between groups may be obvious from such a plot. If the data do not contain distinct natural groups, then one must choose an arbitrary cutoff point that separates the classes. In order to assure objectivity, someone other than the structure–activity person should decide these cutoffs. Usually, the biologist who tests the compounds can do so.

The classifications chosen must be exclusive; that is, no compound can be both active and inactive. Once set, the class limits cannot be changed to get a better discrimination. Ideally, the groupings are exhaustive; no intermediate-potency compounds are omitted. This criterion is relaxed in the case of a somewhat nonreproducible biological test on the basis that it is better to do analysis on reliable data (i.e., definitely active and definitely inactive compounds) than on the basis of some questionable data. If compounds within a certain potency range are not included in the analysis, then the function calculated will technically not apply to compounds of the omitted potency, but estimates from the resulting model can be used to test the model.

V.   LESSONS LEARNED

- The biological endpoint used for QSAR should reflect the logarithm of the rate or binding constant for the compound of interest.
- Including less potent compounds increases the information content of a final QSAR.
- The biological endpoint should be relevant to the goals of the study.

REFERENCES

1. Ariens, E. J.; van Os; G. A. J.; Simonis, A. M.; van Rossom, J. M. In *Molecular Pharmacology: The Mode of Action of Biologically Active Compounds*; Ariens, E. J., Ed. Academic Press: New York, 1964, pp. 136–200, 395–418.
2. Goldstein, A.; Aronow, L.; Kalman, S. M. *Principles of Drug Action: The Basis of Pharmacology*, 2nd ed. Wiley: New York, 1974.
3. Kenakin, T. *Molecular Pharmacology: A Short Course*. Blackwell Science, Oxford, 1997.
4. Brunton, L.; Lazo, J.; Parker, K. *Goodman & Gilman's: The Pharmacological Basis of Therapeutics*, 11th ed. McGraw-Hill: New York, 2006.
5. McEwen, C. M., Jr.; Sasaki, G.; Jones, D. C. *Biochemistry* **1969**, *8*, 3952–62.
6. Collection of terms, symbols, equations, and explanations of common pharmacokinetic and pharmacodynamic parameters and some statistical functions. 2004. http://www.agah-web.de/publikationen_pharma.html (accessed Nov. 15, 2007). Association for Applied Human Pharmacology. Medium: pdf.
7. Riggs, D. S. *The Mathematical Approach to Physiological Problems: A Critical Primer*. Williams & Wilkins: Baltimore, 1963.

6 Form of Equations That Relate Potency to Physical Properties

The preceding chapters have described separately the calculation of the physical and biological properties of molecules. This chapter discusses the form of the equations that relate biological potency to the physical properties of molecules. Many such relationships are not linear correlations with the individual properties, but instead either plateau as a property increases or increase to an optimum and then decrease as the property is further increased. These nonlinear relationships exist because even in the simplest in vitro biological test system there are competing interactions. The competing interactions may arise from the ligands themselves, which often can exist as a mixture of different ionic states, conformers, and tautomers; from the biological system, which may have hydrophobic regions in addition to the target binding site or several aqueous compartments of different pH; or from the interplay between the ligands and the biological system: molecules with different pK_a's would equilibrate differently to biological phases of different pH and may have a different electrostatic affinity for a charged binding site. This chapter will illustrate the use of model-based equations as a method to investigate such complex interrelationships.[1–4]

In contrast to model-based equations, empirical equations are used to fit experimental observations, but the relationships cannot be explained by an explicit model. An example is the parabolic relationship between potency and log P.

In the classic book *The Mathematical Approach to Physiological Problems*, three chapters are devoted to the philosophy and methodology of deriving and checking model-based equations.[5] Precedents for model-based QSAR equations came from the field of pharmacokinetics.[6–10] McFarland,[11] Hyde,[12] Kubinyi,[13] and, later, Buchwald[14,15] proposed models to explain the nonlinear relationship between log P and biological potency. Balaz and coworkers explored models for bioactivity based on log P and ionization or conformational equilibria[16–27] and have also used model-based equations for QSAR analysis.[13,16–18,20,26,28–34]

I.   INTRODUCTION TO MODEL-BASED EQUATIONS

A model-based equation is derived from an idealized hypothesis or compartmental model of the characteristics of the biological system. Such a model hypothesizes the properties of the compartments and the transfer or equilibration of the compounds between the compartments: For example, a model hypothesizes whether a particular compartment is nonaqueous or aqueous. If it is aqueous, the pH of a compartment may be hypothesized or fitted during the analysis. Discrete nonaqueous phases with the same dependence of equilibrium constant on $\log P$ together form one compartment. Similarly, all aqueous phases in equilibrium at the same pH will appear as one compartment in the analysis. Hence, whereas phases are anatomically distinct, compartments differ only in chemical or physical characteristics. The model also postulates how each equilibrium or rate constant depends on the properties of the molecules. The final model-based equation describes the expected relationship between the various molecular properties and potency.

Hence, the models recognize that in a biological system a compound is a dynamic mixture of different ionization states, tautomers, and conformers. Only one, or a very few of these, is the form that binds to the target biomolecule; however, we know which is the bioactive one only if we have definitive structural biology, mechanistic, or special structure–activity information. Model-based equations can provide hypotheses as to which is the bioactive form and provide hypotheses as to why the relationship between physical and biological properties might not be linear.

Although model-based equations provide a powerful strategy for developing a QSAR, a weakness is that a fit to one model does not indicate that it is accurate. In fact, the only information we gain about the model from such fits is that if data do not fit a model-based equation, then the model does not apply to that data. Successful fits may occur by chance, and different models can produce identical equations. Furthermore, model-based equations may be difficult to derive and fit, and many of the equations derived in this chapter contain more unknowns than one has data to support. Although a model-based equation may not be appropriate for a particular set of biological data, an understanding of such equations may be helpful in the interpretation of fits to an empirical equation.

II.   EQUATION FOR AN EQUILIBRIUM MODEL FOR IONIZABLE COMPOUNDS FOR WHICH AFFINITY IS A FUNCTION OF LOG P AND ONLY THE NEUTRAL FORM BINDS

Which $\log P$ should be used in structure–activity calculations of a set of ionizable compounds when pK_a varies within the series? That is, should the calculation use the $\log P$ of the neutral form or the $\log P$ measured at some pH of biological significance? This is an important problem that will serve as an introduction to the philosophy of using a model-based approach to the investigation of a structure–activity relationship. Specifically, we will derive a function of the expected form of the relationship

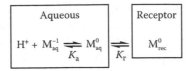

SCHEME 6.1 The equilibrium model for the partitioning of a molecule (M) between the receptor and an aqueous compartment. The molecule also ionizes in the aqueous compartment. The superscripts indicate the charge on the molecule, and the subscripts indicate the compartment.

between log ($1/C$), pK_a, and $\log P$ of the molecules This section will demonstrate the detailed steps in the derivation of a model-based equation. Scheme 6.1 shows the model that we will consider. It describes the simplest in vitro system, one that contains only an aqueous solution and the target biomolecule.

The first assumption is that three forms of each molecule i are in equilibrium: the ionic form in the aqueous compartment (M_{aq}^{-1}); the neutral form in the aqueous compartment (M_{aq}^{0}); and the neutral form in the receptor compartment (M_{rec}^{0}), the compartment that contains the target biomolecule. The first equilibrium constant of interest is the acid dissociation constant K_a. We will assume that the molecule is electrically neutral in the protonated form, as in Scheme 6.1:

$$K_{a,i} = \frac{[H^+] \times \left[M_{aq,i}^{-1} \right]}{\left[M_{aq,i}^{0} \right]}$$
(4.4)

From the definition of the fraction not ionized, ($1 - \alpha$), in Chapter 4:

$$1 - \alpha_i = \frac{\left[M_{aq,i}^{0} \right]}{\left[M_{aq,i}^{0} \right] + \left[M_{aq,i}^{-1} \right]} = \frac{1}{1 + K_{a,i}/[H^+]}$$
(4.7)

and

$$\left[M_{aq,i}^{0} \right] + \left[M_{aq,i}^{-1} \right] = \frac{\left[M_{aq,i}^{0} \right]}{1 - \alpha_i}$$
(6.1)

The second equilibrium constant of interest is K_r, the binding or affinity constant of the molecule for the target biomolecule. It is the ratio of concentration of the neutral form in the receptor compartment to that of the neutral form in the aqueous compartment:

$$K_{r,i} = \frac{\left[M_{rec,i}^{0} \right]}{\left[M_{aq,i}^{0} \right]}$$
(6.2)

The second assumption for this model is that all molecules of the series have identical intrinsic activity. Thus, each requires the same concentration at the target biomolecule to produce an identical biological effect: relative potency is thus proportional to the fraction of the total added compound that is bound to the target. In equation form:

$$\frac{1}{C_i} = X \frac{V_{rec}\left[M^0_{rec,i}\right]}{V_{rec}\left[M^0_{rec,i}\right] + V_{aq}\left(\left[M^0_{aq,i}\right] + \left[M^{-1}_{aq}\right]\right)} \tag{6.3}$$

C_i is the total concentration of molecule i at which relative potency is measured; the V's refer to the volume of the compartment in question; and X is the proportionality constant between amount of compound in the receptor compartment and potency. The numerator and denominator of Equation 6.3 may be divided by $V_{rec}[M^0_{rec,i}]$:

$$\frac{1}{C_i} = X \frac{1}{1 + \dfrac{V_{aq}\left(\left[M^0_{aq,i}\right] + \left[M^{-1}_{aq}\right]\right)}{V_{rec}\left[M^0_{rec,i}\right]}} \tag{6.4}$$

Substituting Equation 6.1 into Equation 6.4:

$$\frac{1}{C_i} = X \frac{1}{1 + \dfrac{V_{aq}\left[M^0_{aq,i}\right]}{V_{rec}\left[M^0_{rec,i}\right] \times (1 - \alpha_i)}} \tag{6.5}$$

K_r may now be substituted for $[M^0_{rec}]/[M^0_{aq}]$:

$$\frac{1}{C_i} = X \frac{1}{1 + \dfrac{V_{aq}}{V_{rec} K_{r,i} (1 - \alpha_i)}} \tag{6.6}$$

The final assumption is that the log of the affinity constant, log K_r, is a linear function of the octanol–water log P:

$$\log K_{r,i} = a' + b \log P_i \text{ or } K_{r,i} = a'' P_i^b \tag{6.7}$$

in which a' equals the logarithm of a''. Equation 6.7 also assumes that the only effect of substituents on binding to the target is a hydrophobic effect. In other words, it assumes that changes in electronic character affect only the fraction of the compound in the neutral form in the various compartments. The effect of relaxing this assumption will be discussed in Section 6.4 of this chapter.

Substituting Equation 6.7 into Equation 6.6 produces the desired relationship:

$$\frac{1}{C_i} = X \cdot \frac{1}{1 + \dfrac{1}{aP_i^b(1-\alpha_i)}} = X \frac{aP_i^b(1-\alpha_i)}{aP_i^b(1-\alpha_i)+1} \tag{6.8}$$

or, in logarithmic terms:

$$\log(1/C_i) = X' + \log(1-\alpha_i) + b\log P_i - \log\left(aP_i^b(1-\alpha_i)+1\right) \tag{6.9}$$

in which a, b, and X or X' for a particular set of data would be fit by nonlinear regression analysis (how to do so will be discussed in Chapters 8 and 10). The constants a and X' are collections of constants; $a = \dfrac{a''V_{rec}}{V_{aq}}$ and $X' = \log X \times a$; and $(1-\alpha)$ is calculated from the pH and pK_a of the compounds, as discussed in Chapter 4, Section I.C.2.

A plot of Equation 6.9 is shown in Figure 6.1. The different lines are for different fractions of the compound ionized. At constant fraction ionized and low log P, the slope of the log $(1/C)$ versus log P plot is b, the slope of the log (K_r) versus log P relationship. This is because at low log P's most of the compound remains in the aqueous compartment, so that the amount in the receptor compartment does not represent a significant fraction of the total: $aP^b \ll 1.0$ and, hence, the final term of Equation 6.9 is zero:

$$\log(1/C_i) = X' + \log(1-\alpha_i) + b\log P_i \tag{6.10}$$

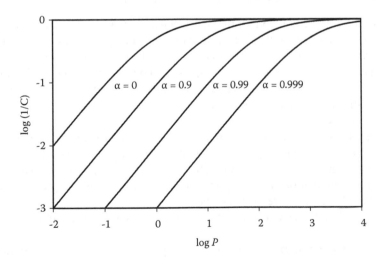

FIGURE 6.1   A plot of Equation 6.9. The constants are $X' = 0.0$ and $a = b = 1.0$. The fraction of compound ionized is indicated on each line.

On the other hand, at high log P most of the compound is bound to the target, and further increases in log P cannot increase this amount. The lines approach an asymptote at a potency equal to X. In this case $aP_i^b(1-\alpha_i) \gg 1.0$, with the result that

$$\log(1/C_i) = X' + \log(1-\alpha_i) + b\log P_i - \log\left(aP_i^b(1-\alpha_i)\right) = X \qquad (6.11)$$

Note that in Equation 6.9 the coefficient of log $(1 - \alpha)$ is not b, but unity. This means that it is not correct to correlate potency with a partition coefficient measured at a pH at which appreciable ionization occurs. The equilibria of ionization and partitioning should be considered separately. According to this model, from Figure 6.1 it can be seen that compounds that are highly ionized but act in the neutral form can show activity provided that the log P of the neutral form is high enough.

An equation of exactly the form as Equation 6.9, except that ionization was not taken into account, was derived from consideration of a diffusion model of transport.[9] Obviously, when more than one model can produce the same equation, one must not think that the model is the only description of reality; rather, it is one possible description.

This simple example shows the power of model-based equations. If a set of structure–activity data can be fit by such an equation, each coefficient can be interpreted directly in terms of the model. Consideration of this simple two-compartment model has led to an equation of a different form from the empirical linear and parabolic ones. It is conceivable that some perfectly reasonable relationships between potency and log P have not been recognized because they are asymptotic rather than linear or parabolic.

III. EQUATIONS FOR EQUILIBRIUM MODELS FOR IONIZABLE COMPOUNDS THAT DIFFER IN TAUTOMERIC OR CONFORMATIONAL DISTRIBUTION AND AFFINITY IS A FUNCTION OF LOG P

With the preceding example in mind, more general equations for equilibrium models may be derived. We will consider the cases in which pK_a varies within the set of analogs with the possibility that the ion as well as the neutral form might bind to the target biomolecule, in which there are lipophilic compartments into which the analogs can partition, and in which there are two aqueous compartments of different pH. Similarly, this section will examine models in which the analogs can exist in variable ratios of tautomers or conformers.

In the models to be considered in this section, it is assumed that log potency, log $(1/C)$, is proportional only to the fraction of the compound that is in the receptor compartment, and that this is not a function of sigma or E_s. Example derivations are not provided, but can be derived by the methods employed in the previous section. Only the final equations are presented here.

The procedure to derive equations for equilibrium models is as follows:

1. All of the equilibrium and mass law expressions that apply to the model are stated. That is, it is assumed that the total amount of the compound is constant: This is generally true of in vitro assays.

2. A general equation for the dependence of the observed biological response on these constants is formulated. It is assumed that concentrations may be used in the equilibrium constants.
3. The relationship between each individual equilibrium constant and physical properties is formulated.
4. The physical properties are substituted into the equation from Step 2 to give biological potency as a function of physical properties.

A. One Aqueous, One Receptor, and One Inert Nonaqueous Compartment, Variable pK_a within the Series

In this model, Scheme 6.2, the compound partitions between an aqueous compartment, a biologically inert nonaqueous compartment, and the receptor compartment. The model assumes that the ionic form of the molecule is not present in the nonaqueous compartment. Such a model might apply to an in vitro test system that contains proteins or lipids, such as found in membranes, in addition to the target biomolecule. The equation for this model is thus

$$\frac{1}{C_i} = X \frac{V_{rec}\left[M^0_{rec,i}\right]}{V_{rec}\left[M^0_{rec,i}\right] + V_{aq}\left(\left[M^0_{aq,i}\right] + \left[H_q M^q_{aq,i}\right]\right) + V_{non}\left[M^0_{non}\right]} \qquad (6.12)$$

The first new assumption for this model is that the logarithm of the equilibrium constant for partitioning the molecule between the aqueous and the nonaqueous compartment (K_n) is proportional to log P:

$$\log K_{n,i} = \log d' + c' \log P_i \qquad (6.13)$$

The model also assumes that the binding constant of the molecule for the target biomolecule is a function of log P (Equation 6.7). If both the ion and the neutral form bind to the target biomolecule, then it is assumed that b, the dependence of binding on log P, is the same for both forms. The two forms might have a different value for a', however.

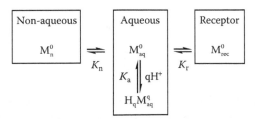

SCHEME 6.2 The equilibrium model for the partitioning of a molecule (M) between the receptor, aqueous, and inert nonaqueous compartments. The molecule ionizes in the aqueous compartment, and the neutral form of the molecules binds to the target biomolecule.

Substituting Equations 6.7 and 6.13 into Equation 6.12, Equation 6.14 is obtained, which relates log $1/C$ to log P and log $(1 - \alpha)$ if only the neutral form of the molecule binds to the target biomolecule:

$$\log(1/C_i) = \log \frac{1}{1 + dP_i^c + \dfrac{1}{aP_i^b(1 - \alpha_i)}} + X \qquad (6.14)$$

A comparison of Equation 6.14 with Equation 6.9 shows that adding another compartment, the nonaqueous compartment, to the model results in the addition of another term in the sum in the denominator, namely, dP_i^c. Recall that the denominator comes from an expression for the total amount of compound in the system. If a compound can be in more compartments, then the number of terms in this sum will increase.

If only the ionic form binds to the target biomolecule, the equation becomes

$$\log(1/C_i) = \log \frac{1}{1 + dP_i^c \dfrac{(1 - \alpha_i)}{\alpha_i} + \dfrac{1}{aP_i^b \alpha_i}} + X \qquad (6.15)$$

If both forms bind to the target biomolecule, it becomes

$$\log(1/C_i) = \log \frac{1 + z \dfrac{\alpha_i}{1 - \alpha_i}}{1 + z \dfrac{\alpha_i}{1 - \alpha_i} + dP_i^c + \dfrac{1}{aP_i^b(1 - \alpha_i)}} + X \qquad (6.16)$$

The values of a, b, c, d, and X of Equations 6.14, 6.15, or 6.16 would be fitted in the nonlinear regression analysis of the structure–activity data. They have the following definitions in terms of the model:

$$a = \frac{a'V_{rec}}{V_{aq}} \qquad (6.17)$$

$$c = c' - b \qquad (6.18)$$

$$d = \frac{d'V_{nonaq}}{a'V_{rec}} \qquad (6.19)$$

$$z = \frac{a'_{ion}}{a'_{neutral}} \qquad (6.20)$$

Hence, a fit to Equations 6.14, 6.15, or 6.16 supplies hypotheses with respect to the relative volumes of the various compartments, the influence of hydrophobicity on binding to the target biomolecule and the inert nonaqueous biological compartment, and the relative contribution to activity of the ionic and neutral form of the molecule.

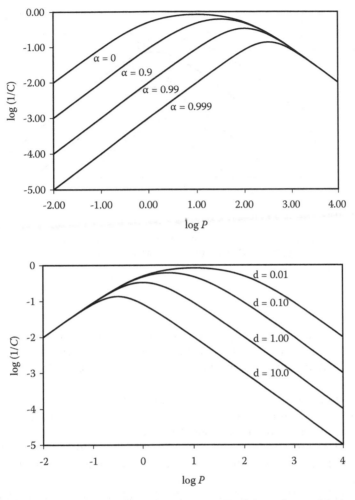

FIGURE 6.2   Plot of Equation 6.14. The constants are $X = 0.0$, $a = b = c = 1.0$ (the top figure is redrawn from Martin, Y. C.; Hackbarth, J. J. *J. Med. Chem.* **1976**, *19*, 1033–9, Figure 3. With permission.) The top graph shows the effect of increasing α, the fraction of compound ionized at $d = 0.01$. The bottom graph shows the effect of increasing d, the relative volume of the inert nonaqueous compartment at $\alpha = 0$.

What is the shape of the relationship between potency and log P according to these equations? What influence do differences in ionization have? Figure 6.2 is a plot of an example for which only the neutral form binds to the target biomolecule (Equation 6.14). The values of the constants were chosen so that the curve would resemble a parabola. Note that with this model, the consequence of increased ionization is to increase the optimum log P and to decrease the potency at the optimum log P. It is seen that as the relative volume of the "inert" nonaqueous compartment increases, both the optimum log P and the potency at the optimum log P decrease.

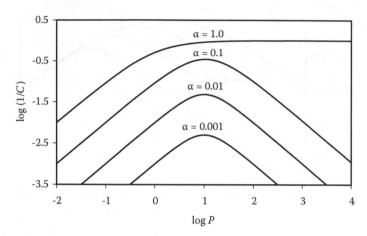

FIGURE 6.3   A plot of Equation 6.15. The constants are $X = 0.0$, $a = b = c = 1.0$, and $d = 0.01$. (Redrawn from Martin, Y. C.; Hackbarth, J. J. *J. Med. Chem.* **1976**, *19*, 1033–9, Figure 4. With permission.)

Figure 6.3 is a similar example for which only the ionic form binds to the target biomolecule (Equation 6.15). The fraction of molecule ionized, α, is indicated on the graph. The influence of ionization on potency is different for Equations 6.14 and 6.15.

Equation 6.16 is particularly complex with respect to ionization. Situations can be imagined in which the shape of the curve is essentially identical to one or the other of the aforementioned functions. However, it can also be used to fit data in which the relationship between potency and log P has two distinct optima depending on whether the ion or the neutral form predominates in the solution of a particular compound.

For Equations 6.14–6.16, at low log P the slope of the log $(1/C)$ versus log P curve is b, the slope of the log P dependence of binding to the target biomolecule. At high log P the slope is $-c$, the difference between the dependence of log P on partitioning to the nonaqueous compartment and affinity for the target biomolecule. In all three equations, only if there is hydrophobic bonding to the target biomolecule will there be a positive slope of log $(1/C)$ versus log P. The only restriction on the slope at high log P is that it must be less than b. It can be positive, zero, or negative. Log $(1/C)$ will be a parabolic-like function of log P only if the dependence of log K_r on log P is half that of the dependence of log K_n on log P. This will be true if the nonaqueous compartment is two-fold more lipophilic than the target biomolecule.[35]

It is not necessary that all terms in Equations 6.14–6.16 be statistically significant. In fact, frequently one term in the sum in the numerator or denominator will not contribute to the value of the sum. For example, if the partitioning to the target biomolecule and to the inert nonaqueous compartments change the same amount with changes in log P, then $c = 0.0$ and the slope of the curve is b at low log P and zero at high log P. This situation is thus identical to the simpler model considered in the previous section. Another simplification occurs in series in which pK_a is constant. In such a case, the $(1 - \alpha)$ and α terms are constant, and fits to Equations 6.14–6.16 cannot be distinguished.

If the amount of compound in the receptor compartment is an insignificant fraction of the total, the 1.0 in the denominator of Equations 6.14–6.16 is absent. In this situation, when c is positive, the curves do not flatten out at maximum potency, but rather, they move up to a peak and abruptly change slope to a downward direction.

As suggested by Figure 6.2 for all cases in which there is an optimum $\log P$, the optimum $\log P$ for Equation 6.14 is not constant, but varies with $\log (1 - \alpha)$. The equation is

$$\log P_{opt} = \frac{1}{b+c}\log\frac{b}{cd} + \frac{1}{b+c}\log\frac{1}{a(1-\alpha)} \tag{6.21}$$

The $\log (1/C)$ at the optimum for Equation 6.14 when $b + c = 1.0$ is

$$\log(1/C_{opt}) = \log\frac{1}{1+dP_{opt}^c + \dfrac{cdP_{opt}^c}{b}} + X \tag{6.22}$$

For Equation 6.15 when $b + c = 1.0$ it is

$$\log(1/C_{opt}) = \log\frac{1}{1+dP_{opt}^c + \dfrac{cdP_{opt}^c(1-\alpha)}{b\alpha}} + X \tag{6.23}$$

B. ONE AQUEOUS, ONE RECEPTOR, AND ONE INERT NONAQUEOUS COMPARTMENT, VARIABLE TAUTOMERIC OR CONFORMATIONAL EQUILIBRIA WITHIN THE SERIES

In this model, Scheme 6.3, again the molecule partitions between an aqueous compartment, a biologically inert nonaqueous compartment, and the receptor compartment. However, in this case there are different forms of the molecule in the inert as well as the aqueous compartments.

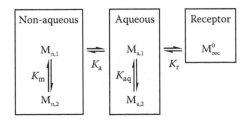

SCHEME 6.3 The equilibrium model for the partitioning of a molecule between tautomeric or conformational states in the nonaqueous and aqueous compartments. Only one tautomer or conformer of the molecule binds to the target biomolecule.

Four equilibrium constants govern this system. The constant K_m describes the equilibrium between the bioactive tautomer, M_1, and another tautomer, M_2, in the nonaqueous phase, and K_{aq} that for the aqueous phase:

$$K_{m,i} = \frac{[M_{n,i,1}]}{[M_{n,i,2}]} \tag{6.24}$$

$$K_{a,i} = \frac{[M_{a,i,1}]}{[M_{n,i,1}]} \tag{6.25}$$

$$K_{aq,i} = \frac{[M_{a,i,1}]}{[M_{a,i,2}]} \tag{6.26}$$

$$K_{r,i} = \frac{[M_{rec,i,1}]}{[M_{a,i,1}]} \tag{6.27}$$

Again, we assume that the equilibrium constant for partitioning the molecule between the aqueous and the nonaqueous compartment (K_n) is proportional to $\log P$ (Equation 6.13), and that partitioning of the molecule to the target biomolecule is also a function of $\log P$ (Equation 6.7).

The equation that relates $\log 1/C$ to $\log P$ and the tautomeric or conformational ratios is

$$\log\left(1/C_i\right) = \log \frac{1}{1 + eP_i^c \dfrac{(K_{m,i}+1)}{K_{m,i}} + \dfrac{f}{P_i^b}\dfrac{(K_{aq,i}+1)}{K_{aq,i}}} + X \tag{6.28}$$

The constants relate to the model as follows:

$$c = c' - b \tag{6.29}$$

$$e = \frac{V_m d}{V_{rec} a} \tag{6.30}$$

$$f = \frac{V_{aq}}{V_{rec} a} \tag{6.31}$$

What is the shape of the relationship between potency and $\log P$ according to these equations? Figures 6.4 and 6.5 show that the tautomeric or conformational equilibrium does not change the shape of the curve or the position of the optimum: It simply changes the $\log 1/C$ at the optimum $\log P$.

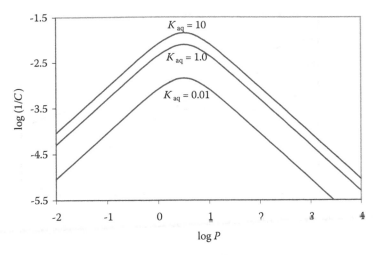

FIGURE 6.4   A plot of Equation 6.28 showing the relationship between potency and log P at various fractions of the active tautomer or conformer. The constants are $X = 0$; $b = c = 0.5$; $e = 10$; $f = 100$; and $K_m = K_{aq}$.

Figure 6.5 shows that when the relative volumes of the receptor, nonaqueous, and aqueous compartments are 1:10:100, then only small increases in potency are realized by increasing the ratio of the bioactive tautomer/conformer to all other tautomers/conformers beyond 10:1 (log $K_{aq} \geq 1.0$). On the other hand, decreasing this ratio decreases potency. Although at a ratio of 10:1 (log $K_{aq} = 1.0$), the decrease in log $(1/C)$ compared to only one form is 0.04, at a ratio of 1:1(log $K_{aq} = 0.0$) the decrease is 0.30; and at a ratio

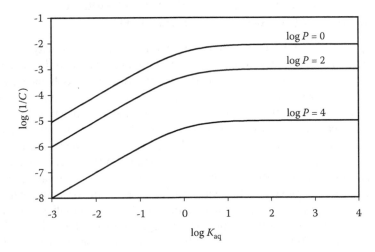

FIGURE 6.5   A plot of Equation 6.28 showing the relationship between potency and the logarithm of the tautomer or conformer ratio at various log P values. The constants are $X = 0$; $b = c = 1.0$; $e = 10$; $f = 100$; and $K_m = K_{aq}$.

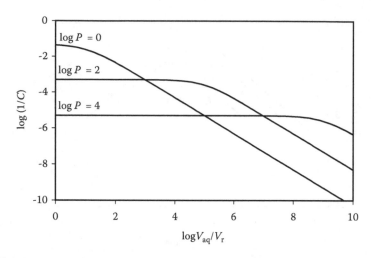

FIGURE 6.6 A plot of Equation 6.28 showing the relationship, at various log P values, between potency and the volume of the aqueous compartment compared to that of the receptor compartment. The constants are $X = 0$; $b = c = 1.0$; $e = 10$; and $K_{m,1} = K_{aq,i} = 1$.

of 1:10 (log $K_{aq} = -1.0$), the decrease is 1.04. Thus, dramatic differences in potency can result if the tautomeric or conformational ratio varies widely within a series.

Log $(1/C)$ is also sensitive to the relative volumes of the various compartments (although these volumes would not vary within one set of assay conditions). Figure 6.6 shows the effect of varying the volume of the aqueous compartment while keeping the volumes of the nonaqueous and receptor compartments constant.

C. ONE RECEPTOR COMPARTMENT AND MULTIPLE NONAQUEOUS AND AQUEOUS COMPARTMENTS OF DIFFERENT pH

Returning to the issue of variable ionization within a structure–activity data set, the more general model includes the equilibration of all ionic and neutral forms between the receptor, multiple nonaqueous compartments of different hydrophobicity, and multiple aqueous compartments of different pH or electrochemical potential. The example with two aqueous, one inert nonaqueous compartment, and the receptor compartment is probably a realistic one for the in vitro antibacterial activity of many substances. In this case, one aqueous compartment would be the solution outside the bacterial membrane, the medium, and the other would be the solution inside the bacterial membrane; and the nonaqueous compartment would be "inert" bacterial proteins and lipids.

Because the complexity of the equation increases with the complexity of the model, this type of model can lead to equations with a large number of terms. One must be certain that a fit to an equation with a large number of terms is justified over a simpler equation. At a minimum, there should be three to five observations for each parameter fit as well as an independent test set.

In this derivation, it is assumed that the concentration of the neutral form is the same in all aqueous compartments. In these equations, there are m inert nonaqueous compartments and n aqueous compartments of different pHs. The compound in the receptor compartment is in equilibrium with the compound in aqueous compartment n. The subscripted coefficients are defined by analogy to those for the m shown in Scheme 6.1.

If it is the neutral form that binds to the target biomolecule:

$$\log(1/C_i) = \log \frac{1}{1 + \sum_{j=1,m}^{m} d_j P_i^{c_j} + \sum_{k=1,n}^{n} \frac{1}{a_k P_i^{b_k}(1-\alpha_{k,i})}} + X \tag{6.32}$$

As noted earlier, it is necessary to use the term for a particular compartment in the statistical fit only if the term corresponding to it contributes significantly to the value of $\log (1/C)$, that is, only if it contributes significantly to the sum in the denominator of Equation 6.32. For example, because the volume of aqueous compartment number 1 usually is the largest, if the cutoff point for significant contribution is 1%, then aqueous compartment i can be ignored if $a_k(1-\alpha_k) = 0.01a_1(1-\alpha_1)$. If we assume that $a_1' = a_k'$, then the cutoff becomes $V_k(1-\alpha_k) = 0.01V_1(1-\alpha_1)$. Thus, both the volume and the fraction not ionized (or pK_a) determine if it is necessary to use a multicompartment model in which only the nonionized form binds to the target biomolecule.

For example, in an assay that measures antibacterial potency by the effect of a compound on microbial growth rate, the maximum bacterial concentration is 10^9/mL.[36] This represents a ratio of V_1/V_2 of approximately 250. On the other hand, in an assay that measures antibacterial potency at stationary phase, the bacterial concentration would be at least 10^{10}/mL, representing a ratio of V_1/V_2 of 25 or less. Clearly, one is more likely to need to consider the second aqueous compartment, the volume of the receptor compartment for the latter type of assay.

If it is the ionic form that binds to the target biomolecule, the equation becomes

$$\log(1/C_i) = \log \frac{\dfrac{\alpha_{n,i}}{1-\alpha_{n,i}}}{\dfrac{\alpha_{n,i}}{1-\alpha_{n,i}} + \sum_{j=1,m}^{m} d_j P_i^{c_j} + \sum_{k=1,n}^{n} \dfrac{1}{a_k P_i^{b_k}(1-\alpha_{k,i})}} + X \tag{6.33}$$

Because of the term $\alpha_n/(1 - \alpha_n)$, it will always be necessary to specifically consider the pH of the nth aqueous compartment. If this pH is not known but the pK_a's of the molecules are known, then the pH can be fitted in the regression analysis.

If both forms bind to the target biomolecule, the equation becomes

$$\log(1/C_i) = \log \frac{1 + z \dfrac{\alpha_{n,i}}{1-\alpha_{n,i}}}{1 + z \dfrac{\alpha_{n,i}}{1-\alpha_{n,i}} + \sum_{j=1,m}^{m} d_j P_i^{c_j} + \sum_{k=1,n}^{n} \dfrac{1}{a_i P_i^{b_k}(1-\alpha_{k,i})}} + X \tag{6.34}$$

The $\log (1/C)$ versus $\log P$ dependencies of Equations 6.32 and 6.33 are similar to those for Equations 6.14 and 6.15. The slope at low $\log P$ is always b, the slope of the $\log (K_{rec})$ versus the $\log P$ relationship. If there is only one inert nonaqueous compartment, then the slope at high $\log P$ is $-c$. If there are several nonaqueous compartments, then the slope of the high $\log P$ portion of the curve would not be constant but would include portions of progressively more negative slope.

Completely different equilibrium models will result in different equations to be fitted. Specifically, if the ionic form of the molecule binds to the nonaqueous compartment, different equations must be used. We have not seen such an instance in our work so far, but it is theoretically possible.

IV. EQUATIONS FOR EQUILIBRIUM MODELS FOR WHICH AFFINITY DEPENDS ON STERIC OR ELECTROSTATIC PROPERTIES IN ADDITION TO LOG *P*

Because other physical properties may also influence affinity for the target biomolecule, in this section Equation 6.16 will be expanded to include other physical properties. Corresponding changes could be made to the other equations, if necessary.

A. AFFINITY FOR THE TARGET BIOMOLECULE IS A FUNCTION OF HYDROPHOBIC INTERACTIONS WITH ONLY CERTAIN POSITIONS OF THE MOLECULES

The binding site on the target biomolecule might not be uniformly hydrophobic, but instead, it might recognize hydrophobic substituents at only certain positions of the molecules. In such a case, the affinity of molecules to the target might not depend on their overall partition coefficients, but rather, on the hydrophobicity, the π values, of the substituents at these positions. The result would be that the $\log P$'s in the aP^b term in the various equations would be replaced by one or more in which P is replaced by the antilog of π values. In a similar fashion, hydrophobic bonding to the inert nonaqueous compartment might be dependent on the hydrophobicity at only certain positions. In this case, the dP^c terms would be modified.

The use of several π values rather than one partition coefficient in a regression analysis increases the number of variables to fit, which increases the possibility of a chance correlation. The general rule of thumb of at least three to five observations per variables fit should be followed. Additionally, an independent test set must be used.

B. POTENCY IS A FUNCTION OF SIGMA OR E_s

If potency is a function not only of the concentration in the receptor compartment, but also steric or electronic effects, then one may simply add the terms $\rho\sigma_i + eE_s$ to the equation to be fitted. Equation 6.16, for example, would then become

$$\log(1/C_i) = \log \frac{1 + z\dfrac{\alpha_i}{1-\alpha_i}}{1 + z\dfrac{\alpha_i}{1-\alpha_i} + dP_i^c + \dfrac{1}{aP_i^b(1-\alpha_i)}} + \rho\sigma_i + eE_{s,i} + X \quad (6.35)$$

When the molecules of interest are ionized in the nonprotonated form, for example, if they are carboxylic acids, then Equation 6.35 can be formulated in terms of the pK_a's of the analogs and the pH of the aqueous compartment:

$$\log(1/C_i) = \log \frac{1 + z\dfrac{K_{a,i}}{[H^+]}}{1 + z\dfrac{K_{a,i}}{[H^+]} + dP_i^c + \dfrac{1 + K_{a,i}}{aP_i^b[H^+]}} + pK_{a,i} + eE_{s,i} + X \qquad (6.36)$$

If the molecules of interest are instead ionized in the protonated form, then the $K_a/[H^+]$ term would be inverted.

C. AFFINITY FOR THE TARGET BIOMOLECULE IS A FUNCTION OF SIGMA OR E_s

If the equilibrium constant for binding to the target biomolecule is a function of steric or electrostatic properties, then Equation 6.7 could be modified as follows:

$$\log K_{r,i} = \log a' + b \log P_i + \rho\sigma_i + eE_{s,i} \qquad (6.37)$$

and Equation 6.16 would become

$$\log(1/C_i) = \log \frac{1 + z\dfrac{\alpha_i}{1 - \alpha_i}}{1 + z\dfrac{\alpha_i}{1 - \alpha_i} + dP_i^c + \dfrac{1}{10^{\rho\sigma_i}10^{eE_{s,i}}aP_i^b(1 - \alpha_i)}} + X \qquad (6.38)$$

D. PARTITIONING TO THE INERT NONAQUEOUS COMPARTMENT IS A FUNCTION OF SIGMA OR E_s

If, however, it is the equilibrium constant to the inert nonaqueous compartment that is a function of steric and electrostatic properties, then Equation 6.16 becomes

$$\log(1/C_i) = \log \frac{1 + z\dfrac{\alpha_i}{1 - \alpha_i}}{1 + z\dfrac{\alpha_i}{1 - \alpha_i} + 10^{\rho\sigma_i}10^{eE_{s,i}}dP_i^c + \dfrac{1}{aP_i^b(1 - \alpha_i)}} + X \qquad (6.39)$$

Hence, σ or E_s will enter the equation in different ways, depending on the process that is affected. In particular, the electronic effect of substituents can be divided into their separate influence on ionization and on the strength of the various types of interactions.

TABLE 6.1
Examples of Simplifications of Equation 6.36: Both Forms Bind and Affinity Depends on pK_a, Ignoring the E_s Term[4]

Dominant Compartment	Only the Neutral Form Binds	Only the Ion Binds
Receptor	$\log(1/C_i) = (g \times pK_{a,i}) + X$	$\log(1/C_i) = (g \times pK_{a,i}) + X$
Nonaqueous	$\log(1/C_i) = X - (d \times \log P_i) + (g \times pK_{a,i})$	$\log(1/C_i) = X + \mathrm{pH} + ((g-1)pK_{a,i}) - (d \times \log P_i)$ substituting $X' = X + \mathrm{pH}$ and $g' = g - 1$: $\log(1/C_i) = X' - (d \times \log P_i) + (g' \times pK_{a,i})$
Aqueous	$\log(1/C_i) = X - \log\left(1 + \dfrac{K_{a,i}}{[H^+]}\right) + (b \times \log P_i) + (g \times pK_{a,i})$	$\log(1/C_i) = X - \log\left(1 + \dfrac{[H^+]}{K_{a,i}}\right) + (b \times \log P_i) + (g \times pK_{a,i})$

Source: Martin, Y. C. In *Physical Chemical Properties of Drugs*; Yalkowsky, S. H., Sinkula, A. A., Valvani, S. C., Eds. Marcel Dekker: New York, 1980, pp. 49–110.

E.   SIMPLIFICATION OF THE MODELS IF ONE OR MORE TERMS IS NOT SIGNIFICANT

Although a model-based equation may contain many terms, one or more terms may not contribute significantly if for all the analogs the concentration of a particular form in some compartment is negligible compared to the total concentration. Table 6.1 shows some of the simplifications of Equation 6.36 that can occur. Table 6.1 also shows that different models can also lead to the same equations.

V.   EQUATIONS FOR MODELS THAT INCLUDE EQUILIBRIA AND THE RATES OF BIOLOGICAL PROCESSES

Frequently, one wishes to correlate the observed rate of a biological transfer, absorption from the gut, for example, with the physical proportion of the molecules involved. A typical model for such a situation is diagrammed in Scheme 6.4. There are two equilibria prior to a rate-limiting step for the actual transfer. The first equilibrium is that with a proton to form the ionic and neutral form. The second is equilibration of the neutral form between the aqueous compartment and the membrane surrounding it. The slow step is transfer of the compound from the membrane into another compartment indicated by the rate constant k_b.

A reasonable assumption is that the logarithm of the aqueous membrane equilibrium constant, $\log K$, is a linear function of $\log P$:

$$\log K_{m,i} = i + h \log P_i \tag{6.40}$$

A further assumption is that the logarithm for the rate constant for the transfer out of the membrane, $\log k_b$, is also a linear function of $\log P$:

$$\log k_{b,i} = n + j \log P_i \tag{6.41}$$

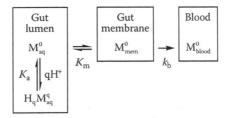

SCHEME 6.4   Model for the gastrointestinal absorption of ionizable substances. Two equilibria are established: the ionization equilibria in the lumen of the gut, and equilibration of the neutral form of the molecule between the lumen and the membrane surrounding the gut. The slow step in absorption is assumed to be movement of molecules out of this membrane into the blood. The superscripts indicate the charge on the molecule and the subscripts the compartment.

From these assumptions and material balance considerations, Equation 6.42 may be derived:

$$\log k_i = \log \frac{1}{dP_i^c + \dfrac{1}{aP_i^b(1-\alpha_i)}} + X \tag{6.42}$$

As previously stated, a, b, c, and d are adjustable constants to be fitted by nonlinear regression analysis. These constants have a meaning based on the model: a and d are functions of the intercepts of the two physical-biological property relationships and the volumes of the various compartments, $b = j + h$, and $c = h$. In terms of observed structure–activity relationships, b is the slope of the log k versus log P function at low log P, and $-c$ is the slope at high log P.

If the individual partition or rate constants depend on physical properties other than log P, then these factors would appear in the first or second term in the denominator of Equation 6.41, depending on whether the prior equilibrium or the rate-limiting step, respectively, was affected.

For a series in which the degree of ionization does not vary, Kubinyi has proposed that one use Equation 6.43:

$$\log(1/C) = a\log P_i - b\log(\beta P_i + 1) + c \tag{6.43}$$

Note the similar form of this equation to Equation 6.9. Equation 6.43 was derived from the model shown in Scheme 6.5, which is a multicompartment model of alternating aqueous and nonaqueous phases. The compound moves through the system as through a countercurrent apparatus: that is, equilibrium is established between two phases, then all of one type of phase is advanced one position and equilibrium is reestablished. It is assumed that the receptor phase may be of different lipophilicity from the other nonaqueous phases. The coefficients b and c are functions of the lipophilicity of the nonaqueous and receptor phases and the number of phases transversed. The coefficient β is the ratio of the average volume of the lipid to that of the aqueous phases. Several groups have reported fits to Equation 6.24.[13,37–42]

An article by Balaz summarizes the derivation of detailed kinetic models that contain many aqueous and nonaqueous phases.[25] They are applied to the problem of pharmacokinetics: the effect of changes in molecular properties on absorption, distribution, metabolism, and elimination rate or equilibrium constants. The models include the effect of changes in hydrophobicity, ionization, ion pairing, and protein binding.

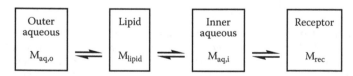

SCHEME 6.5 The countercurrent model for the distribution of molecules from the outer aqueous compartment to the receptor compartment.

VI.   EQUATIONS FOR WHOLE-ANIMAL TESTS FOR WHICH NO MODEL CAN BE POSTULATED

For in vivo tests, a model must consider not only partitioning to the receptor compartment and efficacy in producing a response, but also what proportion of the administered compound is actually in the animal at the time of measurement of the biological response. Thus, the structure–activity relationships of all of the processes of absorption, distribution, metabolism, and excretion (ADME) of the compounds must be considered. This is obviously complex, and leads to an equation with a large number of constants to be fitted. In terms of model-based equations, probably the best that one can do is to use one of the foregoing equations and assume that these ADME processes are either constant within the series or that their net effect is a linear function of the physical properties of interest.

VII.   EMPIRICAL EQUATIONS

The classic structure–activity equation for nonlinear relationships between potency and log P is the parabolic one. Because the use of a parabola is empirical, there is some ambiguity in how ionization should be treated. Two different equations have been used. The first one, Equation 6.44, treats ionization as a modification to be applied to log P:

$$\log(1/C_i) = a + b(\log P_i(1-\alpha_i)) + c(\log P_i(1-\alpha_i))^2 \tag{6.44}$$

A sample plot of this equation is shown in Figure 6.7. Increasing ionization changes the slope of the log P relationship, but changes the potency at the optimum only slightly. In addition, the curves are no longer symmetric. For example, when only 20% of the compound is in the nonionized form ($\alpha = 0.8$), the slope at high log P is almost zero although is it larger at low log P.

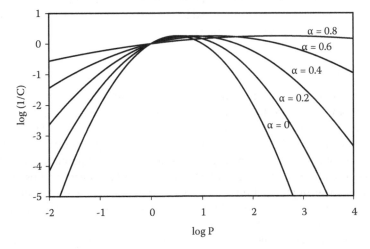

FIGURE 6.7   A plot of Equation 6.44. The constants are $a = 0.0$ and $b = c = 1.0$.

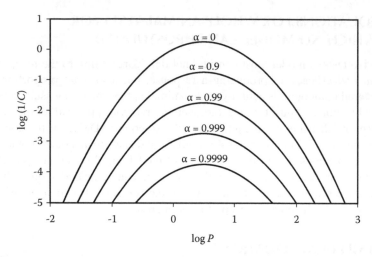

FIGURE 6.8   A plot of Equation 6.45. The constants are $a = 0.0$ and $b = c = 1.0$. (Redrawn from Martin, Y. C.; Hackbarth, J. J. *J. Med. Chem.* **1976**, *19*, 1033–39, Figure 5. With permission.)

The second approach to the influence of ionization is to assume that potency should be corrected to reflect not the total concentration, but only that of the neutral form. In equation form this becomes

$$\log(1/C_i) - \log(1 - \alpha_i) = a + b\log P_i + c(\log P_i)^2 \tag{6.45}$$

A plot of Equation 6.42 is shown in Figure 6.8. In this case the influence of ionization is to not change the optimum log P, but rather to decrease the potency at the optimum. If one assumes that the ionic form is responsible for activity, one would substitute $\log(\alpha_i)$ for the $\log(1 - \alpha_i)$ term.

It should be clear that neither of the two empirical equations predicts the same relationships between potency, log P, and fraction ionized over the whole log P range, as do the several model-based equations.

VIII.   LESSONS LEARNED

This chapter shows that there are many possible relationships, linear and nonlinear, between a physical property and biological potency of a series of molecules. How does one decide which is the one to be used for a particular data set? This will be discussed statistically in Chapters 7 and 8, but from the foregoing discussion it is clear that plots of the data can be revealing. For example, a plot of potency versus log P of all points of relatively constant pK_a and tautomeric and conformer ratios may be linear, linear to an asymptote, or linear with a positive slope at low log P and linear with a lower positive or negative slope at high log P. These cases would be properly fitted with equations of the form of Equations 6.7, 6.8, and 6.16, 6.32, 6.42, or 6.43, respectively. In certain circumstances, one may choose to use the empirical Equations 6.44 or 6.45. The influence of ionization and tautomeric or conformational equilibria may also be discerned from plots. The graphs presented here may be a guide. Hence, the approach to be used is not to immediately try to fit every possible equation, but rather to try to decide which equation seems most likely to be correct.

Several examples were presented to show that different models could lead to the same equation. Conversely, the same model with different assumptions can lead to different equations. As a consequence, the models must not be taken literally unless other experiments prove their applicability. Philosophically, even though the equations are model based, frequently the decision as to which to use with a particular data set is decided empirically.

REFERENCES

1. Martin, Y. C.; Hackbarth, J. J. *J. Med. Chem.* **1976**, *19*, 1033–9.
2. Martin, Y. C.; Hackbarth, J. J. In *Chemometrics: Theory and Application*; Kowalski, B., Ed. American Chemical Society: Washington, DC, 1977, pp. 153–64.
3. Martin, Y. C. In *Quantitative Structure–Activity Analysis*; Franke, R., Oehme, P., Eds. Akademie-Verlag: Berlin, 1978, pp. 351–8.
4. Martin, Y. C. In *Physical Chemical Properties of Drugs*; Yalkowsky, S. H., Sinkula, A. A., Valvani, S. C., Eds. Marcel Dekker: New York, 1980, pp. 49–110.
5. Riggs, D. S. *The Mathematical Approach to Physiological Problems: A Critical Primer.* Williams and Wilkins: Baltimore, 1963.
6. Higuchi, T.; Davis, S. S. *J. Pharm. Sci.* **1970**, *59*, 1376–83.
7. Flynn, G. L.; Yalkowsky, S. H. *J. Pharm. Sci.* **1972**, *61*, 838–52.
8. Wagner, J. G.; Sedman, A. J. *J. Pharmacokin. Biopharm.* **1973**, *1*, 23–50.
9. Yalkowsky, S. H.; Flynn, G. I. *J. Pharm. Sci.* **1973**, *62*, 210–7.
10. Flynn, G. L.; Yalkowsky, S. H.; Weiner, N. D. *J. Pharm. Sci.* **1974**, *63*, 300–4.
11. McFarland, J. W. *J. Med. Chem.* **1970**, *13*, 1192–6.
12. Hyde, R. M. *J. Med. Chem.* **1975**, *18*, 231–3.
13. Kubinyi, H. *J. Med. Chem.* **1977**, *20*, 625–9.
14. Buchwald, P. *J. Pharm. Sci.* **2005**, *94*, 2355–79.
15. Buchwald, P. *Math. Biosci.* **2007**, *205*, 108–36.
16. Balaz, S. *Quant. Struct.-Act. Relat.* **1994**, *13*, 381–92.
17. Balaz, S.; Pirselova, K.; Schultz, T. W.; Hermens, J. *J. Theor. Biol.* **1996**, *178*, 7–16.
18. Pirselova, K.; Balaz, S.; Schultz, T. W. *Arch. Environ. Contam. Toxicol.* **1996**, *30*, 170–7.
19. Pirselova, K.; Balaz, S.; Ujhelyova, R.; Sturdik, E.; Veverka, M.; Uher, M.; Brtko, J. *Quant. Struct.-Act. Relat.* **1996**, *15*, 87–93.
20. Dvorsky, R.; Balaz, S.; Sawchuk, R. J. *J. Theor. Biol.* **1997**, *185*, 213–22.
21. Jantova, S.; Balaz, S.; Stankovsky, S.; Spirkova, K.; Lukacova, V. *Folia Biologica* **1997**, *43*, 83–9.
22. Pirselova, K.; Balaz, S.; Sturdik, E.; Ujhelyova, R.; Veverka, M.; Uher, M.; Brtko, J. *Quant. Struct.-Act. Relat.* **1997**, *16*, 283–9.
23. Hornak, V.; Balaz, S.; Schaper, K. J.; Seydel, J. K. *Quant. Struct.-Act. Relat.* **1998**, *17*, 427–36.
24. Balaz, S.; Lukacova, V. *Quant. Struct.-Act. Relat.* **1999**, *18*, 361–8.
25. Balaz, S. *Am. J. Pharm. Ed.* **2002**, *66*, 66–71.
26. Lukacova, V.; Balaz, S. *J. Chem. Inf. Comput. Sci.* **2003**, *43*, 2093–105.
27. Khandelwal, A.; Lukacova, V.; Kroll, D. M.; Raha, S.; Comez, D.; Balaz, S. *J. Phys. Chem. A* **2005**, *109*, 6387–91.
28. Potts, R.; Guy, R. *Pharm. Res.* **1995**, *12*, 1628–33.

29. Ishizaki, J.; Yokogawa, K.; Nakashima, E.; Ichimura, F. *J. Pharm. Pharmacol.* **1997**, *49*, 762–7.
30. Ishizaki, J.; Yokogawa, K.; Nakashima, E.; Ichimura, F. *J. Pharm. Pharmacol.* **1997**, *49*, 768–72.
31. Balaz, S. *Perspect. Drug Discovery Des.* **2000**, *19*, 157–177.
32. Poulin, P.; Theil, F. P. *J. Pharm. Sci.* **2000**, *89*, 16–35.
33. Poulin, P.; Schoenlein, K.; Theil, F. P. *J. Pharm. Sci.* **2001**, *90*, 436–47.
34. Buchwald, P.; Bodor, N. *J. Med. Chem.* **2006**, *49*, 883–91.
35. Leo, A.; Hansch, C. *J. Org. Chem.* **1971**, *36*, 1539–44.
36. Garrett, E. R.; Mielck, J. B.; Seydel, J. K.; Kessler, H. J. *J. Med. Chem.* **1969**, *12*, 740–5.
37. Kubinyi, H. *QSAR: Hansch Analysis and Related Approaches*. VCH: Weinheim, 1993, Vol. 1.
38. Devillers, J.; Bintein, S.; Domine, D. *Chemosphere* **1996**, *33*, 1047–65.
39. Mor, M.; Zani, F.; Mazza, P.; Silva, C.; Bordi, F.; Morini, G.; Plazzi, P. V. *Farmaco* **1996**, *51*, 493–501.
40. Verma, R. R.; Hansch, C. *ChemBioChem* **2004**, *5*, 1188–95.
41. Fujikawa, M.; Nakao, K.; Shimizu, R.; Akamatsu, M. *Bioorg. Med. Chem.* **2007**, *15*, 3756–67.
42. Itokawa, D.; Nishioka, T.; Fukushima, J.; Yasuda, T.; Yamauchi, A.; Chuman, H. *QSAR Comb. Sci.* **2007**, *26*, 828–36.

7 Statistical Basis of Regression and Partial Least-Squares Analysis

This chapter will develop the statistical background necessary for quantitative structure–activity studies on modest-size data sets of related molecules, usually no more than a hundred analogs. Extension of this background to larger, more diverse data sets will be discussed in Chapter 11. The discussions that follow will develop linear regression analysis in some detail.[1,2] Once it is understood, it can be generalized to other multivariate statistical methods, in particular, partial least-squares (PLS) regression.

I. FUNDAMENTAL CONCEPTS OF STATISTICS

A. DEFINITIONS

Every biological and chemical measurement has an associated error. Statistical methods are used to provide estimates of the variable quantities as well as the expected error of the estimates. Such methods also evaluate the significance of apparent differences between sets of observations.

1. A *population* is the totality of observations about which inferences are to be made. A *sample* is a subset of a population.
2. The *arithmetic mean* of a sample is the sum of the values of all sample observations divided by the number of observations, n. In statistical notation:

$$\overline{X} = \frac{\sum_{i=1}^{n} X_i}{n} = \frac{\sum X_i}{n} \tag{7.1}$$

In this equation $\overline{X}$ (X bar) is the mean; Σ refers to the operation of summation. The small letters above and below the summation symbol (n and i) in the left equation refer to the values of X to be included in the sum; that is, all i's from X_i, the first, to X_n the nth. Since all X's are included, these indices may be omitted as in the right-hand equation.

3. The *residual* or *deviation* (dev) is the difference between the observed and calculated values of a property. For example, the deviation from the mean is the difference between the observed value and the mean.
4. The *sum of squares of deviations* from the mean (SS$_{mean}$) is calculated by squaring each deviation and then summing these values:

$$SS_{mean} = \sum (\bar{X} - X)^2 \tag{7.2}$$

5. The *degrees of freedom* (*DF*) is a number associated with any statistical quantity. It indicates how many independent pieces of information regarding the *n* numbers $X_1, X_2, X_3 \ldots X_n$ are needed to form the statistic. The SS$_{mean}$(*n* − 1) degrees of freedom because the mean and the preceding (*n* − 1) values determine the *n*th value.
6. An estimate of the *variance* of a population (s^2_{mean}) is the sum of squares of deviations from the sample mean, divided by the degrees of freedom:

$$s^2_{mean} = \frac{SS_{mean}}{n-1} = \frac{\sum (\bar{X} - X)^2}{n-1} \tag{7.3}$$

This quantity forms the basis for very useful statistical calculations called *analysis of variance,* which will be discussed briefly below.
7. An estimate of the *standard deviation* of a population (s_{mean}) is a commonly used measure of dispersion or scatter of the observation from the mean. It is the square root of the variance:

$$s_{mean} = \sqrt{\frac{\sum (\bar{X} - X)^2}{n-1}} \tag{7.4}$$

8. A *normal distribution* of the values of some property *Y* is observed if there are many additive and independent factors that contribute to the value of *Y*. The mean of a normal distribution is μ; and the standard deviation is σ. In QSAR it is usually assumed that the residuals, the difference between the observed and the predicted potencies of the analogs are normally distributed.

For normally distributed populations, the mean plus or minus one standard deviation includes 68.3% of the observations; plus or minus two standard deviations, 95.5%; and plus or minus three standard deviations, 99.7%.

As an example of the calculations described by Equations 7.1–7.4, consider the data set in Table 7.1.[3] It lists the name, log *P*, and log (1/C) (log molar potency) of a set of compounds that inhibit the swelling of fibrin. Figure 7.1 shows the relationship between log *P* and log (1/C). The sums of the log *P* values and the log (1/C) values are listed below the individual numbers. The mean log *P* is the sum (4.39) divided by the number of observations (8), or 0.55; the mean log (1/C) is 11.25/8 = 1.41.

TABLE 7.1

Sample Calculation of Statistical Properties (example V-80[3])

	1	2	3	4	5	6	7	8
Name	log P	$dev^2_{log P}$	log (1/C)	$dev^2_{log(1/C)}$	$dev_{log P} \times dev_{log(1/C)}$	Pred. log(1/C)	Residual	(Residual)^2
Chloroform	1.97	2.02	2.80	1.93	1.97	2.55	0.25	0.0625
Methanol	−0.66	1.46	0.90	0.26	0.62	0.43	0.47	0.221
Ethanol	−0.16	0.50	0.90	0.26	0.36	0.84	0.06	0.004
Propanol	0.34	0.04	0.60	0.66	0.17	1.24	−0.64	0.407
Butanol	0.88	0.11	1.51	0.01	0.033	1.67	−0.16	0.027
Pentanol	1.40	0.72	2.02	0.37	0.52	2.09	−0.073	0.005
Ethyl ether	0.77	0.05	1.92	0.26	0.11	1.58	0.34	0.112
Ethyl carbamate	−0.15	0.49	0.60	0.66	0.57	0.84	−0.24	0.059
Calculations								
Sum	4.39	5.39	11.25	4.41	4.35			0.898
DF	8	7	8	7				6
Sum/DF	0.549	0.77	1.41	0.63				0.15
$\sqrt{Sum/DF}$		0.88		0.79				0.39

The calculation of the variance and standard deviation of log P is shown in Column 2, and those of log (1/C) in Column 4. First, the mean is subtracted from each observation to calculate a deviation, and then the deviation is squared. The sum of these squared deviations is SS_{mean}. In the example, SS_{mean} for log P is 5.39; for log (1/C) it is 4.41. Using Equation 7.3, the variance of log P is 5.39/7, or 0.77 and for

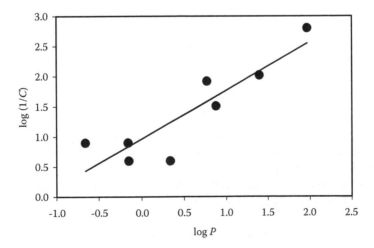

FIGURE 7.1 A plot of the data from Table 7.1. The least-squares line calculated from Equation 7.17 is also shown.

log $(1/C)$ it is 4.41/7, or 0.63. The corresponding standard deviations (Equation 7.4) are the square roots of these numbers: 0.88 and 0.79, respectively.

B. PROBABILITY, ESTIMATION, AND HYPOTHESIS TESTING

How does one decide if observed differences between samples are likely or unlikely to occur by chance? How does one establish the expected confidence in a prediction?

From the definition of the normal distribution, 95.5% of the observations of a sample of X's with a mean value of μ_x will fall within $\mu_x \pm 2\sigma_{\bar{x}}$, where $\sigma_{\bar{x}}^2 = \sigma^2/n$. In statistical notation, the 95.5% confidence interval of μ_x is $\mu_x \pm 2\sigma_{\bar{x}}$.

However, if the sample size is small, one does not have a reliable estimate of $\sigma_{\bar{x}}$. In such a case, it is necessary to use the Student's t distribution to calculate confidence intervals. Student's t values are available in printed tables or included in computer programs such as Microsoft Excel. The t value depends on the number of degrees of freedom and the desired probability level, α, which is calculated from the desired probability of confidence P as $1 - P$. For example, a 95% confidence interval would be calculated with α equal to 0.05.

Equation 7.5 is the formula to calculate the confidence intervals for the mean using the Student's t values:

$$\hat{\bar{X}} = \bar{X} \pm t_{\alpha,n-1} \sqrt{1/n} \times s_{mean} \qquad (7.5)$$

$\hat{\bar{X}}$ (X bar, hat) gives the expected upper and lower limits of $\bar{X}$. For example, 2.365 is the Student's t value for seven degrees of freedom and 95% probability; hence, the 95% confidence limits of the mean log$(1/C)$ values in Table 7.1 are $1.41 \pm (2.365 \times 0.354 \times 0.793) = 1.41 \pm 0.66$ (or 0.75 to 2.07).

The corresponding limits of a future individual observation X are calculated using Equation 7.6:

$$\hat{\bar{X}} = X \pm t_{\alpha,n-1} \sqrt{1+1/n} \times s_{mean} \qquad (7.6)$$

For the example, the 95% confidence limits of the log$(1/C)$ of an individual value are 1.41 ± 1.98.

C. ANALYSIS OF VARIANCE (ANOVA)

Analysis of variance tests whether two or more groups of samples could have been obtained from the same population. It partitions the total sum of squares of deviation from the overall mean (SS$_{mean}$) into (1) the sum of squares of deviation within each group from its mean (SS$_w$), plus (2) the sum of squares of deviation of group means from the overall mean (SS$_b$). In equation form:

$$SS_{mean} = SS_w + SS_b = \sum (\bar{\bar{X}} - X)^2 = \sum (\bar{X} - X)^2 + \sum (\bar{\bar{X}} - \bar{X})^2 \qquad (7.7)$$

$\bar{\bar{X}}$ (X, double bar) is the overall mean; $\bar{X}$ the mean of an individual group.

TABLE 7.2

General ANOVA Table for Difference between Group Means

(g groups, m observations per group)

Source of Variation	Sum of Squares (SS)	Degrees of Freedom (DF)	Mean Square (MS)
Between groups	$SS_b = \sum (\bar{X} - \bar{\bar{X}})^2$	$g - 1$	$\sum \dfrac{(\bar{X} - \bar{\bar{X}})^2}{g-1} = MS_b$
Within groups	$SS_w = \sum (X - \bar{X})^2$	$g(m - 1)$	$\sum \dfrac{(X - \bar{X})^2}{g(m-1)} = MS_w$
Total about mean	$SS_{mean} \sum (X - \bar{X})^2$	$gm - 1$	$\sum \dfrac{(X - \bar{\bar{X}})^2}{gm-1} = s^2_{mean}$

There are three different degrees of freedom (DF) for ANOVA: The total DF about the mean is the number of observations minus 1, that for variation within groups is the group size minus one, and that for variation between means is the number of groups minus 1. Table 7.2 shows a typical setup for ANOVA using the symbols g, the number of groups; and m, the number of observations in each group (assumed to be of equal size for the example).

The *F distribution* is used to test the hypothesis that two samples came from a population of identical mean. The F ratio to test the difference in means is the ratio of the mean square deviation between groups (MS_b) to the mean square within groups (MS_w). For the case of g groups of size m, the relationship is

$$F_{g-1,g(m-1)} = \frac{MS_b}{MS_w} = \frac{\sum (\bar{\bar{X}} - \bar{X})^2/(g-1)}{\sum (X - X)^2 (g(m-1))} \qquad (7.8)$$

As the MS_b becomes smaller, the sample means of the individual groups become closer and the F ratio at constant MS_w also becomes smaller. To test the significance of the result, one compares the sample F to theoretical F of the appropriate probability. As with t values, F values are available in printed tables or built in as functions in computer programs such as Microsoft Excel. If the calculated ratio exceeds the theoretical one, then the difference between the means is statistically significantly different.

II. SIMPLE LINEAR REGRESSION

In quantitative structure–activity studies, the aim is to estimate the direction and magnitude of the change in potency that will result from the change of a given molecular property. In the simplest case, the log potency Y is a linear function of one predictor property X only. Although these calculations are usually performed with a computer, this section will discuss the calculation and statistical evaluation of such a straight-line relationship. This discussion of simple linear regression forms the basis for the discussion of multiple regression in the next section.

A. ASSUMPTIONS

1. The values of the independent or predictor property X (log P, for example) are measured or calculated without error.
2. The expected mean values for the dependent property Y (log $(1/C)$, for example), μ_Y, for all values of X lie on a straight line:

$$\mu_Y = \beta_0 + \beta_1 X \tag{7.9}$$

3. For any given value of X, the Y's are independently and normally distributed. This assumption states that E in Equation 7.10 is a normally distributed error term with a mean of zero:

$$Y = \beta_0 + \beta_1 X + E \tag{7.10}$$

4. The variance around the regression line is constant; that is, independent of the magnitude of X and Y.

B. LEAST-SQUARES LINE CALCULATION

Recall that in the analysis of variance, Equation 7.7, we partitioned SS_{mean} into SS_b and SS_w. In a similar manner, for a regression equation we partition SS_{mean} into that due to regression, SS_{reg}, and that due to error, SS_{error}.

$$SS_{mean} = SS_{regr} + SS_{error} \tag{7.11}$$

The sum of squares due to regression comes about because, according to the linear relationship, at values of X below $\overline{X}$ we expect Y to be some value above or below $\overline{Y}$, and at values of X above $\overline{X}$ we expect Y to be some value in the opposite direction from $\overline{Y}$. SS_{error} is simply the sum of squares of the deviations (or residuals) of observations from the regression line. It is calculated using values from Equation 7.10 as the predictions:

$$SS_{error} = \sum E_i^2 = \sum (Y_i - \beta_0 - \beta_1 X_i)^2 = \sum (Y_i - \hat{Y})^2 \tag{7.12}$$

Note that the predicted value is indicated by the symbol $\hat{Y}$ (Y hat). Because the best line minimizes SS_{error} with respect to β_0 and β_1, we differentiate SS_{error} with respect to β_0 and β_1, set the derivatives equal to zero, solve these equations, and use b's rather than β's to denote the change from the true coefficients to the best estimate of them. The resulting equations are

$$b_1 = \frac{\sum (X_i - \bar{X}) \times (Y_i - \bar{Y})}{\sum (X_i - \bar{X})^2} \qquad (7.13)$$

and

$$b_0 = \bar{Y} - b_1 \bar{X} \qquad (7.14)$$

Consider again the data in Table 7.1. The sum of the cross products in the numerator of Equation 7.14 is 4.35 (Column 6), and the sum of squares in the denominator (Column 3) is 5.39. Hence, the slope of the regression line is

$$b_1 = 4.35/5.39 = 0.807 \qquad (7.15)$$

and the intercept is

$$b_0 = 1.41 - (0.807 \times 0.549) = 0.967 \qquad (7.16)$$

Thus, the least-squares equation for this data is

$$\log(1/C) = 0.967 + 0.807 \log P \qquad (7.17)$$

This line is shown in Figure 7.1.

C. SIGNIFICANCE OF THE OBSERVED RELATIONSHIP: F AND R^2

Recall from Equation 7.11 that the total sum of squares of deviation about the mean can be partitioned into that due to regression and that due to error. As data fit a line better and better, the SS_{error} becomes smaller and smaller. The R^2 statistic takes advantage of this property:

$$R^2 = \frac{SS_{regr}}{SS_{mean}} = \frac{SS_{mean} - SS_{error}}{SS_{mean}} \qquad (7.18)$$

Thus, as SS_{error} becomes smaller, R^2 approaches 1.0; as it becomes larger, R^2 approaches 0.0. Equation 7.18 also shows that R^2 is the fraction of the total sum of squares of deviations from the mean that is explained by the regression equation.

To calculate R^2 for the example in Table 7.1, we first calculate the predicted potency, $\log(1/C_{calc})$, for each observation by substituting the values in Column 2 into Equation 7.17 to produce the values in Column 6, which we then use to calculate the residual

TABLE 7.3
General ANOVA Table for Simple Regression (*n* observations, 1 predictor property)

Source of Variation	Sum of Squares (SS)	Degrees of Freedom (DF)	Mean Square (MS)
Regression	$SS_{regr} = \sum (\hat{Y} - \bar{Y})^2$	1	$\sum (\hat{Y} - \bar{Y})^2 = MS_{regr}$
About regression	$SS_{error} = \sum (Y - \hat{Y})^2$	$n - 2$	$s^2 = MS_{error} = \dfrac{\sum (Y - \hat{Y})^2}{n - 2}$
About mean	$SS_{mean} = \sum (Y - \bar{Y})^2$	$n - 1$	$s^2_{mean} = \dfrac{\sum (Y - \hat{Y})^2}{n - 1}$

$(\log(1/C)_{obs} - \log(1/C)_{calc})$ shown in Column 8 and its square, Column 9. R^2 is calculated from the sum of these values and the sum of squares deviation from the mean:

$$R^2 = (4.41 - 0.898)/4.41 = 0.796 \qquad (7.19)$$

Thus, only 79.6% of the variation in potency is explained by variation of log *P*. Inspection of Figure 7.1 reveals that, indeed, not all points fall on the calculated line.

Table 7.3 shows an ANOVA table for simple regression analysis. The calculated *F* ratio is

$$F_{1,n-2} = MS_{regr}/MS_{error} \qquad (7.20)$$

based on 1 (one predictor was used) and $n - 2$ (the number of observations, minus 1 for the mean and one for the regression coefficient) degrees of freedom. It is used to assess the statistical significance of the observed regression.

The ANOVA calculations for the example in Table 7.1 are shown in Table 7.4. The *F* ratio calculated from the data is

$$F_{1,6} = 3.51/0.15 = 23.4 \qquad (7.21)$$

TABLE 7.4
ANOVA Table for Example V-80

Source of Variation	Sum of Squares (SS)	Degrees of Freedom (DF)	Mean Square (MS)
Regression	$SS_{regr} = 4.41 - 0.90$ $= 3.51$	1	$MS_{regr} = 3.51$
About regression (residual)	$SS_{error} = 0.90$	6	$s^2 = MS_{error} = 0.15$
About mean	$SS_{mean} = 4.41$	7	$s^2_{mean} = 0.63$

From tables, the calculated probability of exceeding this F ratio by chance is 0.0029. Hence, the slope of the line is highly statistically significant from zero.

Because R^2 and F are computed from related quantities, they, too, are related:

$$F = (n-2)\frac{R^2}{1-R^2} \tag{7.22}$$

For the example, this is

$$F = 6(0.796/0.204) = 23.6 \tag{7.23}$$

An alternative method may be used to assess the statistical significance of the calculated regression parameter b_1. It consists of calculating the Student's t value:

$$t = \frac{b_1\sqrt{(X_i - X)^2}}{s} \tag{7.24}$$

If t is greater than that tabulated at the required confidence level and $n-2$ degrees of freedom, then the slope of the line is significantly different from zero. The significance of the intercept (i.e., is $b_0 = 0.0?$) can be assessed in a parallel manner. For simple regression, the t value calculated by Equation 7.24 is the square root of the F ratio calculated by Equations 7.20 or 7.22.

D.   CHARACTERISTICS OF s, THE STANDARD ERROR OF ESTIMATE

In Table 7.3 the residual mean square deviation from the regression line was designated s^2. Its square root, s, is the root mean square deviation or standard error of estimate. It measures how precisely the calculated line fits individual values. It is used in the calculation of the confidence interval of a predicted value. The equation is

$$\hat{Y}_k \pm t_{\alpha, n-2} \times s \times \sqrt{1 + \frac{1}{n} + \frac{\sum (X_k - \bar{X})^2}{SS_{mean}}} \tag{7.25}$$

Thus, the precision with which the expected potency of a new compound may be estimated is directly proportional to the magnitude of s.

A related question is, what is gained in precision of estimate by using a regression equation rather than the mean to predict a new observation? In order for such a procedure to be sensible, s should be "substantially" less than s_{mean}. In the example from Table 7.1, $s = 0.149$, whereas $s_{mean} = 0.63$. Hence, a substantial improvement in precision is gained by use of the regression equation rather than the mean.

If there is some prior estimate of the variance of individual replicated measurements, then one can compare this value with s^2. If the two are approximately equal, then the regression model fits the data as precisely as could be expected. The larger the ratio $s_{mean}/s_{replicate}$, the more likely one is to establish a statistically significant relationship.

Sometimes the regression relationship is not statistically significant but $s^2 = s^2_{replicate}$. This means that the overall variation in the response was not large enough to permit discovery of a significant relationship with the predictor property. The solution to this problem is to expand the range of observations of Y. In structure–activity problems, this may occur if the relevant physical property has been varied over too narrow a range or if the potency of less potent analogs has not been established. In such a case, the potency would also vary only slightly, and it would be impossible to prove a relationship between the property and potency.

To give a feel for other properties of s, some simple relationships will be calculated with specific numbers:

- What s_{mean} will be required to produce an equation of $R^2 = 0.80$ if $s = 0.10$ and $n = 10$? Assume that an equation for which $s^2 = s^2_{replicate}$ will be found.

$$R^2 = \frac{SS_{mean} - SS_{error}}{SS_{mean}} = 0.80 \tag{7.26}$$

$$SS_{error} = s^2(n-2) = 0.10^2 \times (10-2) = 0.08 \tag{7.27}$$

$$0.80 = \frac{SS_{mean} - 0.08}{SS_{mean}} \tag{7.28}$$

$$SS_{mean} = 0.40 \tag{7.29}$$

$$s^2_{mean} = 0.40/9 = 0.044 \tag{7.30}$$

$$s_{mean} = 0.21 \tag{7.31}$$

Thus, a standard deviation about the mean 2.1-fold greater than the standard deviation of replicates would be required to be able to anticipate achieving a regression equation of $R^2 = 0.80$ if $n = 10$.

- What is the F ratio for the above regression? From Equation 7.20:

$$F_{1,8} = (n-2)\frac{R^2}{1-R^2} = 8\frac{0.80}{0.20} = 32 \tag{7.32}$$

- What is the s_{mean} necessary to anticipate a statistically significant (at the 95% confidence level) relationship if $s = 0.1$ and $n = 10$? Note that $MS_{error} = s^2 = 0.01$, and the F value for 95% confidence at 1 and 8 degrees of freedom is 5.32.

$$F_{1,8} = 5.32 = \frac{SS_{regr}}{MS_{error}} = \frac{SS_{regr}}{0.01} \tag{7.33}$$

$$SS_{regr} = 0.0532 \qquad (7.34)$$

$$SS_{mean} = SS_{regr} + SS_{error} = 0.0532 + (0.010 \times 8) = 0.1332 \qquad (7.35)$$

$$s^2_{mean} = 0.1332/9 = 0.0148 \qquad (7.36)$$

$$s_{mean} = 0.12 \qquad (7.37)$$

- What is the R^2 for the above example?

$$R^2 = \frac{SS_{regr}}{SS_{mean}} = \frac{0.0532}{0.1332} = 0.40 \qquad (7.38)$$

It is up to the user to decide if such a relationship, which describes only 40% of the variance in the data, would be useful.

E. CORRELATION AND REGRESSION

A regression analysis is performed to quantitate the dependence of one property on one or several others. For example, regression analysis in structure–activity studies might examine the dependence of relative potency on the partition coefficient. In contrast, correlation analysis quantitates the interdependence or relatedness of two properties such as octanol–water log P calculated with different programs. The calculation of the correlation coefficient and statistical tests of its significance are identical to those of regression analysis with the exception that the direction of the relationship is maintained as the sign of R.

The interpretation of the meaning of the correlation coefficient differs from regression to correlation. Properties that are correlated are said to covary. Correlation coefficients are used in structure–activity studies to assess the degree of relatedness of the predictor properties used for regression analysis. Highly correlated properties do not provide independent information for the regression analysis.

III. MULTIPLE LINEAR REGRESSION (MLR or OLS)

The relative potency of molecules is often influenced by several physical properties. With multiple regression analysis, one uses these p physical properties to calculate the best least-squares equation to fit the data and then evaluates the significance of the contribution of each physical property to the prediction.

In addition to the assumptions that underlie simple regression, multiple regression analysis also assumes that each predictor property provides independent information.

TABLE 7.5

ANOVA Table for Multiple Regression (*n* observations, *p* predictor properties)

Source of Variation	Sum of Squares (SS)	Degrees of Freedom (DF)	Mean Square (MS)
Regression	$SS_{regr} = \sum (\hat{Y} - \bar{Y})^2$	p	$MS_{regr} = \dfrac{\sum (\hat{Y} - \bar{Y})^2}{p}$
About regression	$SS_{error} = \sum (Y - \hat{Y})^2$	$n - p - 1$	$s^2 = MS_{error} = \dfrac{\sum (Y - \hat{Y})^2}{n - p - 1}$
About mean	$SS_{mean} = \sum (Y - \bar{Y})^2$	$n - 1$	$s^2_{mean} = \dfrac{\sum (Y - \hat{Y})^2}{n - 1}$

That is, that the predictor properties are not correlated with each other. (If some of them are correlated, then a different method of analysis must be used: for example, principal components or partial least-squares analysis. These will be described in the next section.)

The calculation of the coefficients of the properties for the multiple regression equation is identical to that for the simple least-squares line with the exception that b_1, X_i, and $\bar{X}$ of Equations 7.12 and 7.13 refer to vectors (a list) of predictor properties rather than only a single value. The vector X_i, for example, contains the value for each physical property of molecule i, and the vector $\bar{X}$ contains the mean value of each physical property.

Hence, R^2 is calculated from Equation 7.18 with the SS_{regr} evaluated from the multivariate equation. The overall variance ratio, F, is calculated from a generalized ANOVA table, Table 7.5:

$$F_{p,\, n-p-1} = \frac{MS_{regr}}{MS_{error}} \tag{7.39}$$

Thus, the F is associated with the number of predictor properties, p, and the number of observations, n. The relationship between F and the multiple regression R^2 now also involves p. The relationship is

$$F_{p,\, n-p-1} = \frac{n - p - 1}{p} \frac{R^2}{1 - R^2} \tag{7.40}$$

Two F tests are used to test if the contribution of a predictor, the b, is different from zero—that is, significant. The first, the sequential F test, measures the significance of adding the least significant term to the regression equation. It is calculated as the ratio of the increase in the SS_{regr} due to adding the final term to that from the more complex equation. The increase in SS_{regr} is evaluated:

$$SS_j = \Delta SS_{regr} = \left(R_p^2 - R_{p-1}^2\right)SS_{mean} \tag{7.41}$$

Thus,

$$F_{1,n-p-1} = \frac{\Delta SS_{regr}}{MS_{error}} = \frac{\left(R_p^2 - R_{p-1}^2\right)SS_{mean}(n-p-1)}{\left(1 - R_p^2\right)SS_{mean}} = \frac{\left(R_p^2 - R_{p-1}^2\right) \times (n-p-1)}{\left(1 - R_p^2\right)} \tag{7.42}$$

This F has 1 and $n - p - 1$ degrees of freedom. It must be statistically significant for the last property added to the equation, usually that which contributes least to the reduction of the sum of squares. The other test is the partial F test. It is calculated as the sequential F test for each property in the model, but evaluated as if it were added last to the equation. Each of these must be significant.

The confidence interval for a predicted individual observation is a generalized form of Equation 7.25:

$$\hat{Y}_k \pm t_{\alpha,n-p-1} \times s \times \sqrt{1 + \frac{1}{n} + \frac{\sum(X_k - \bar{X})^2}{SS_{mean}}} \tag{7.43}$$

in which X_k is the vector of physical properties of observation k, and $\bar{X}$ is the vector of the mean of the physical properties in the sample.

If one chooses to calculate a fit to Equation 6.44 or 6.45 or a similar equation, one includes the physical property and its square in the multiple regression calculation. This calculates a parabola. A negative coefficient of the squared term indicates a maximum potency at some intermediate value of the physical property; a positive coefficient indicates a minimum. The value of the maximum or minimum, X_0, is calculated by setting the partial derivative of the equation with respect to that property to zero and solving:

Equation:

$$\log 1/C = a + bX + cX^2 \tag{7.44}$$

Derivative:

$$\frac{\partial[\log(1/C)]}{\partial_x} = b + 2cX_0 = 0 \tag{7.45}$$

Thus,

$$X_0 = -b/2c \tag{7.46}$$

IV.   NONLINEAR REGRESSION ANALYSIS (NLR)

One must use nonlinear regression analysis to fit equations that describe a more complex relationship between the dependent variable and the predictor properties.[4] An example is Equation 6.9:

$$\log(1/C_i) = X' + \log(1 - \alpha_i) + b \log P_i - \log\left(aP_i^b(1-\alpha_i)+1\right) \qquad (6.9)$$

As in linear regression, the best estimate of the coefficients is that in which the sum of squares of the difference between observed and calculated potency (SS_{error}) is a minimum, which, from Equation 7.18, also maximizes R^2. However, in nonlinear regression the derivative of SS_{error} with respect to the coefficients a and b is a complex function of these coefficients. The result is that one cannot explicitly solve for them. Instead, one must use one of several iterative procedures that are available.[2,5,6] Although most work well, different programs will often produce similar but not necessarily identical results.

The statistics of a nonlinear regression analysis fit are calculated with the same equations as are used for multiple linear regression. Sections in Chapters 8 and 10 contain further discussion of the challenges and advantages of fitting nonlinear equations.

V.   PRINCIPAL COMPONENTS ANALYSIS (PCA)

If the values of the predictor molecular properties are correlated, then a fundamental assumption of multiple regression analysis is violated. Such correlations can lead to different regression equations in which one or another of the correlated properties is statistically significant but no equation in which both are significant. Principal components and partial least-squares analysis capitalize on the correlations between predictor properties to calculate a smaller number of independent composite properties that contain contributions from each of the original properties. The difference between the two is that principal components are calculated from the predictor properties only, whereas partial least squares considers both the predictor properties and the biological potency. PCA and PLS are presented in detail by the earliest practitioners[7] of PLS and others.[8,9]

The first stage of a PCA is to calculate the correlation matrix, the square root of R^2 calculated as with Equation 7.18, between each pair of predictor properties. Consider Table 7.6, which lists several calculated properties for the molecules considered in Table 4.10.[10–16] Table 7.7 is the correlation matrix of these properties. Note the high correlations of the calculated log solubility with the two calculated log P's, log D, CMR, molecular weight, and number of rotatable bonds. For multiple regression analysis one would consider only one of a pair of highly correlated properties. Note also that the calculated log (Brain/blood) is not highly correlated with any one property, although it is moderately correlated with PSA and log P. (Recall that it was calculated from the CLOGP and PSA values using the equation from Clark.[12])

PCA uses matrix algebra to analyze the correlation matrix. The algorithm is based on the fact that correlated properties lie on a line in the multidimensional space described by the properties. The calculations reveal these lines as orthogonal, perpendicular, principal components or eigenvectors that have been extracted to maximize the variance explained. Hence, each eigenvector represents a different line

TABLE 7.6
Calculated Properties of Molecules from Table 4.10

Molecule	Structure	Solubility Log M[10]	KowWin Log P[11]	Log D pH 7.4[14]	CLOGP Log P[15]	CMR[15]	MW	Rotatable Bonds[16]	Log (Brain/blood)[12]	PSA[13]
Amoxil	4.15	−1.58	−1.36	−3.93	−1.87	9.30	365.44	4	−2.11	133
Keflex	4.16	−1.53	−1.93	−3.95	−1.84	9.20	347.42	4	−1.81	113
Clavulanic acid	4.17	0.70	−2.04	−4.40	−1.06	4.41	199.18	2	−1.31	87
Glucophage	4.18	0.89	−1.40	−3.99	−1.45	3.52	129.2	2	−1.44	91
Glucophage tautomer 1	4.19	0.89	−2.64	−4.31	−1.63	3.52	129.2	3	−1.43	89
Glucophage tautomer 2	4.2	0.89	−1.40	−3.99	−1.63	3.52	129.2	2	−1.46	91
Motrin	4.21	−3.55	3.79	0.38	3.68	6.12	206.31	4	0.15	37
Celebrex	4.22	−4.95	3.47	2.90	4.37	9.14	381.4	3	−0.35	78
Lipitor	4.23	−8.70	6.36	2.22	4.46	15.67	558.7	12	−0.84	112
Viagra	4.24	−4.62	2.30	4.05	2.22	12.58	474.64	7	−1.20	113
Viagra tautomer 1	4.25	−3.02	1.60	2.70	3.56	12.52	474.64	7	−1.00	114
Viagra tautomer 2	4.26	−4.77	2.47	4.05	1.98	12.59	474.64	7	−1.24	113
Claritin	4.27	−7.93	5.66	4.94	5.05	10.83	382.92	3	0.28	42
Biaxin[a]	4.28	−6.36	3.18	1.16	2.37	19.39	748.08	8	−2.21	183

[a] Biaxin is 6–0–methylerythromycin A.

TABLE 7.7
Correlation Matrix of the Properties in Table 7.6

Property	Log M Solubility	KowWin Log P	Log D pH 7.4	CLOGP Log P	CMR	MW	Rotatable bonds	Log (Brain/blood)	PSA
Log M Solubility	1.000								
KowWin Log P	-0.935	1.000							
Log D pH 7.4	-0.828	0.847	1.000						
CLOGP Log P	-0.842	0.942	0.890	1.000					
CMR	-0.873	0.695	0.706	0.612	1.000				
MW	-0.850	0.662	0.695	0.594	0.995	1.000			
Rotatable bonds	-0.871	0.763	0.688	0.647	0.871	0.826	1.000		
Log (Brain/blood)	-0.258	0.529	0.434	0.650	-0.175	-0.204	0.024	1.000	
PSA	-0.336	0.059	0.142	-0.053	0.721	0.744	0.486	-0.793	1.000

through the p-dimensional data space, and each eigenvector describes the contribution (loading) of each original property to that component. One uses these loadings to calculate the principal component scores of each observation.

Although the algorithm produces as many components as there are original properties, usually only a few explain most of the variance of the data. Each component has an associated eigenvalue, λ, that describes the fraction of the overall variance that it explains. The relative magnitudes of the eigenvalues provide information with respect to the number of truly independent properties in a data set. The fraction of the total variance accounted for by λ_i is

$$FV_i = \frac{\lambda_i}{\sum_j \lambda_j}$$

(7.47)

in which FV_i is the fraction of the variance explained by λ_i, and $\Sigma\lambda_j$ is the sum of all λ values. The sum of the eigenvalues of a correlation matrix is equal to the number of properties p. In a data set with no correlation between properties, all λ values are equal and $FV_i = 1.0/p$. However, FV_i can vary from p (all properties are highly correlated) to 0.0 (all variance is explained by other eigenvalues). The rank of the correlation matrix is the number of nonzero eigenvalues. It can be no larger than the smaller of the number of properties or the number of observations.

Although there can be many components with nonzero eigenvalues, there is no clear-cut answer as to how many of these are important. One viewpoint is that the true number of dimensions in a data set is equal to the number of eigenvalues that have a value greater than or equal to 1.0 (i.e., that each explains the variance of at least one independent property). However, if one plans to use the components as descriptors in a regression analysis, then one may choose to include sufficient components to explain 95% of the property variance.

Table 7.8 lists the eigenvalues of the correlation matrix in Table 7.7. Note that the first two eigenvalues explain 94% of the variance in the molecular properties.

TABLE 7.8
Eigenvalues of the Principal Components of the Correlations in Table 7.7

Component	Eigenvalue	Cumulative Variance Explained
1	5.683	0.631
2	2.777	0.940
3	0.278	0.971
4	0.149	0.987
5	0.068	0.995
6	0.037	0.999
7	0.006	1.000
8	0.002	1.000
9	0.000	1.000

TABLE 7.9

Loading of Properties on the Two Largest Principal Components from Table 7.8

	PC1	PC2
Log M Solubility	−0.97	0.09
KowWin Log P	0.93	−0.32
Log D pH 7.4	0.90	−0.26
CLOGP Log P	0.89	−0.42
CMR	0.88	0.45
MW	0.85	0.49
Rotatable bonds	0.79	0.41
Log (Brain/blood)	0.36	−0.93
PSA	0.18	0.97

Table 7.9 and Figure 7.2 show the property loadings on these two principal components. On Figure 7.2, note the close relationship of the log P calculators and also of the bulk calculators molecular weight, CMR and number of rotatable bonds. In contrast, the calculated log Brain/blood, polar surface area, and calculated log solubility are unique. Although polar surface area and log (Brain/blood) have similar loadings on Component 1, they have opposite loadings on Component 2. Also, log solubility has negative loadings on Component 1, whereas the log P and bulk calculators have positive loadings.

Table 7.10 shows the principal component scores of the molecules, and Figure 7.3, a plot. The various tautomers of Glucophage plot closely to each other,

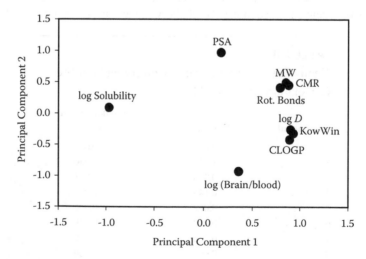

FIGURE 7.2 A plot of the loadings of the molecular properties on the two largest principal components of the data in Table 7.6.

TABLE 7.10

Component Scores of the Molecules from Table 7.6

	PC1	PC2
Amoxil	−0.655	1.099
Keflex	−0.698	0.769
Clavulanic acid	−1.142	−0.113
Glucophage	−1.214	−0.121
Glucophage tautomer 1	−1.259	−0.034
Glucophage tautomer 2	−1.226	−0.102
Motrin	0.100	−1.727
Celebrex	0.487	−0.992
Lipitor	1.577	0.263
Viagra	0.667	0.276
Viagra tautomer 1	0.581	0.189
Viagra tautomer 2	0.667	0.299
Claritin	0.962	−1.761
Biaxin	1.153	1.955

as do the tautomers of Viagra. The two beta-lactam antibiotics Amoxil and Keflex are also close on the plot. Such PC scores can be used as descriptors in a regression analysis.

Figures 7.2 and 7.3 illustrate how a PCA reduces a data table into plots that highlight the relationships between properties or between molecules. Chapter 11 compares the insight that can be gained from plots derived from PCA and other multidimensional projection methods.

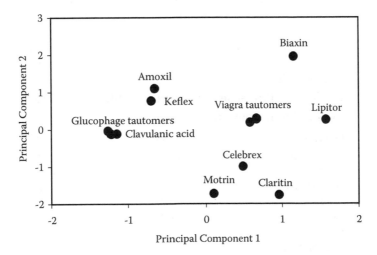

FIGURE 7.3 A plot of the scores of the molecules on the two largest principal components of the data in Table 7.6.

VI. PARTIAL LEAST-SQUARES ANALYSIS (PLS)

Partial least-squares regression combines features of multiple squares regression, sometimes called OLS, with those of principal components.[7,9] In essence, PLS calculates the loadings of the molecular properties on the components, called *latent variables*, to explain the dependent property such as biological potency, not to maximize the explained variance in physical properties as in PCA. As in PCA, each successive latent variable is orthogonal to all previous latent variables. In contrast to PCA, however, in PLS each successive latent variable also explains more and more of the variance in the dependent property, resulting in an increase in the fitted R^2.

The statistical significance of the R^2 for a PLS fit is evaluated with Equation 7.39 using the number of latent variables included as the p degrees of freedom. Although the significance of each successive latent variable can also be evaluated with partial F-tests, usually the number of latent variables to include is chosen from the results of leave-one-out cross-validation, discussed in the following section.

Tables 7.11 and 7.12 summarize the PLS analysis to predict log Brain/blood from the other properties in the sample data. Figure 7.4 shows the loadings of the properties on the two largest latent variables. Note that the properties that are close in

TABLE 7.11
Statistics of PLS Fit of Log (Brain/blood) from Table 7.6

Latent Variable	R^2	s_{fit}	q^2	s_{cv}
1	0.818	0.319	0.595	0.476
2	0.987	0.087	0.976	0.116
3	0.993	0.064	0.984	0.094
4	0.999	0.023	0.989	0.079
5	1.000	0.009	0.998	0.036
6	1.000	0.003	1.000	0.009
7	1.000	0.002	1.000	0.006
8	1.000	0.002	1.000	0.006

TABLE 7.12
Loadings of Properties on Latent Variables of PLS Log (Brain/blood) from Table 7.6

Property	LV1	LV2
Log M Solubility	−0.84	1.49
KowWin Log P	0.98	−1.01
Log D pH 7.4	0.92	−1.08
CLOGP Log P	1.02	−0.79
CMR	0.35	−2.29
MW	0.30	−2.31
Rotatable bonds	0.30	−2.06
PSA	−0.60	−2.01

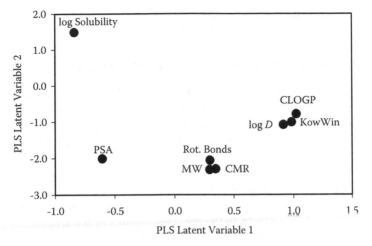

FIGURE 7.4 A plot of the loadings of the molecular properties on the two largest PLS latent variables of the data in Table 7.6.

principal component space (Figure 7.2) are close on this plot also. However, the relationships between these clusters of properties have changed somewhat.

The plot of observed versus the PLS-calculated log (Brain/blood) in Figure 7.5 confirms the fit of the PLS model. The model has no physical meaning because the log (Brain/blood) was calculated from an equation that includes only log P calculated with CLOGP and polar surface area.[12] If PLS had captured the true relationship, only two latent variables would be recovered.

PLS is the statistical method used to generate CoMFA 3D-QSAR models such as are discussed in Chapters 9 and 10. PLS calculates the positive and negative contribution to

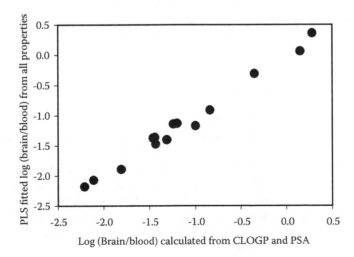

FIGURE 7.5 A plot of the PLS fit using two latent variables of the calculated log (Brain/blood) data in Table 7.6.

potency of the energies at the various lattice points that surround the molecules. These contributions are then displayed as contour plots in the vicinity of the molecules.

VII. ESTIMATING THE PREDICTIVITY OF A MODEL

It was stressed in the descriptions of the various methods that a fundamental assumption is that the values of the predictor properties are known without error. In fact, the variation in calculated log P values in Table 7.6 remind us that the property calculations are not precise. Hence, one may choose to pay special attention to the effect of error in these properties on the reliability of models from the data.[17]

Usually, an important goal of statistical modeling of structure–activity relationships is to forecast the potency of as-yet-untested molecules. Hence, one would like to have a measure—in addition to the confidence limits of the predictions—of how reliable the predictions might be.[17] This topic will be revisited in subsequent chapters; however, the following techniques are standard procedures in QSAR.

A. THE ROLE OF STATISTICS

As implied above, a fundamental criterion is to be sure that all terms in a regression equation are statistically significant or that one has chosen only latent variables that explain a large percentage of the variance in the potency of the molecules in the model. This is only the first step, however.

A fundamental assumption of regression analysis and its derivative, PLS, is that the residuals from the model are normally distributed. Hence, one should examine the residuals to verify that this is the case for the model in question. If a few molecules have much larger residuals that the others, the model might not apply to them.

Subsequent chapters will illustrate how one can select from a pool of possible predictors those that can be used to generate a statistically significant model. However, if one examines too many potential predictors, there is a possibility of chance correlation.[18] There is no statistical measure of this possibility; rather, one must design investigations to make chance correlation unlikely.

B. METHODS THAT TEST A MODEL USING EXTERNAL DATA

1. Separate Training and Test Sets

The most robust method to gain an estimate of the predictivity is to hold back part of the data as a prediction set; to develop a model on the remaining molecules, the training set; and to use the developed model to forecast the potency of the molecules in the prediction set. The molecules of the prediction set may be selected at random or they may also be selected as every n^{th} molecule in the data set sorted by potency. Because the test set must contain more than a few molecules, dividing the data into training and test sets can be used only for data sets that contain at least 20 molecules.

Yet another strategy is to divide the data into three parts: the training set, a validation set to use to decide between alternative models, and a stand-alone test set that is not used in any model development or selection.

Note that one must be certain that the test (and validation) set does not contain molecules that are essentially duplicates of molecules in the training set. The potency of such "twins" will be accurately forecast because this is essentially a reproduction of the fitted values. Rather, the molecules used to test a model must correspond to a different pattern of predictor properties than those used to train the model.[19,20]

2. Cross-Validation

In cross-validation, one develops multiple models for which one or more molecules are omitted. Each model is then used to predict the potency of the omitted molecules. The overall predictivity of the model is then reported as q^2, which is formally equivalent to R^2 except that the predicted values, rather than the fitted values, are used in Equation 7.18.

Leave-one-out (LOO) cross-validation is often used to select how many latent variables to include in a PLS analysis. It is performed by constructing as many models as there are observations, each model differing in the observation omitted. If there are fewer rows than columns, as in CoMFA, row-based calculations are faster than those based on columns.[8] This is the basis of the SAMPLS[21] algorithm that is part of Sybyl CoMFA.

A better strategy is to perform many cross-validations, leaving out a larger fraction of the data. For example, one could repeatedly use randomization to divide the data in half, develop a model from each half, and use the opposing model to estimate q^2. For these methods, one might test hundreds or even thousands of cross-validation models. Yet another strategy is to divide the data into three parts: the training set, the cross-validation set, and a stand-alone test set. The cross-validation set would be used to select the predictors to include in the analysis.

As with dividing data into training and test sets, one must be certain that "twin" molecules are deleted from the analysis.

3. Y-Scrambling

Y-scrambling, also called *response permutation*, removes the correspondence between the potency and predictor properties of the molecules but keeps the relationships between the individual predictor properties. A Y-scrambling procedure randomly shuffles the potency values while not changing the matrix of predictor properties. Usually hundreds to thousands of such Y-scrambled models are generated. Because the correlation between potency and the predictors is removed by the Y-scrambling, no significant model should result. The R^2 and q^2 of any model from the original data should be substantially larger than that from corresponding Y-scrambled models. An advantage of Y-scrambling is that it can be used on small as well as large data sets.

In progressive scrambling, the data is divided into several blocks of observations that fall into a similar range of biological potency. Scrambling is then performed within the blocks.[20]

VIII.   LESSONS LEARNED

The sum of squares of some property is a fundamental concept in statistics. It is based on the difference between two numbers of interest. For example, to calculate the sum of squares of deviation from the mean of some property (SS_{mean}), one calculates the mean, subtracts the mean from each number, squares this value for each item, and then sums these values.

Regression analysis calculates a fit between some dependent property, such as biological potency, and independent properties, such as molecular properties. For regression analysis, the SS_{mean} is partitioned into that due to the fit to the line, SS_{regr}, and that due to error of the fit, SS_{error}. These values are used to calculate the R^2 or variance explained of the fit:

$$R^2 = \frac{SS_{regr}}{SS_{mean}} = \frac{SS_{mean} - SS_{error}}{SS_{mean}} \tag{7.18}$$

Principal components analysis exploits the correlations between properties to highlight the relationships between them and to provide estimates of the true number of independent properties. Each principal component is a linear combination of the original properties, the loadings, and each observation is associated with scores calculated from each component. Plots of the loadings visualize the relationships between the properties, and plots of the scores visualize the relationships between the original observations.

Partial least-squares analysis exploits the correlations between molecular properties to model the relationship between them and the dependent property such as biological potency. It extracts latent variables that are linear combinations of the molecular properties to successively explain more of the dependent property. Plots of the loadings visualize both the relationships between properties and the relationship between these properties and potency. Plots of the scores visualize the relationships between the original observations.

Although statistical significance is a prerequisite for a valid model, one must use other methods to support a model. Typical methods include dividing the data into training and test sets, generating cross-validation statistics, and scrambling the dependent variable to assess the baseline statistics for a random correlation.

REFERENCES

1. Daniel, C.; Wood, F. S. *Fitting Equations to Data: Computer Analysis of Multifactor Data*, 2nd ed. Wiley: New York, 1980.
2. Draper, N. R.; Smith, H. *Applied Regression Analysis*, 3rd ed. Wiley: New York, 1998.
3. Hansch, C.; Dunn III, W. J. *J. Pharm. Sci.* **1972**, *61*, 1–19.
4. Seber, G. A. F.; Wild, C. J. *Nonlinear Regression*. Wiley-IEEE: New York, 2003.
5. Metzler, C. M. *A Brief Introduction to Least Squares Estimation,* 1970 Compilation of Symposia Papers, 5th National Meeting of the APHA Academy of Pharmaceutical Sciences, 380–96.
6. *Nonlinear Estimation;* 2008. http://www.statsoft.com/textbook/stathome.html?stnonlin.html&1 (accessed Feb. 6, 2008), StatSoft, Inc.
7. Eriksson, L.; Johansson, E.; Kettaneh-Wold, N.; Trygg, J.; Wikström, C.; Wold, S. *Multi- and Megavariate Data Analysis. Part I: Basic Principles and Applications*, 2nd ed. Umetrics AB: Umeå, 2007.

8. Bush, B. L.; Nachbar Jr., R. B. *J. Comput.-Aided Mol. Des.* **1993**, *7*, 587–619.
9. Garson, G. D., *Partial Least Squares Regression PLS);* 2008. http://faculty.chass.ncsu.edu/garson/PA765/pls.htm (accessed Nov. 6, 2008).
10. WsKow, version 1.1. Syracuse Research Corporation. Syracuse. 1999.
11. LogKow, version 1.1. Syracuse Research Corporation. Syracuse. 1999.
12. Clark, D. E. *J. Pharm. Sci.* **1999**, *88*, 815–21.
13. Ertl, P.; Rohde, B.; Selzer, P. *J. Med. Chem.* **2000**, *43*, 3714–7.
14. ADME Batch, Pharma Algorithms. Toronto. 2004.
15. CLOGP, version 4.3. Biobyte Corporation. Claremont, CA. 2007. http://biobyte.com/bb/prod/40manual.pdf (accessed Nov. 12, 2007).
16. Martin, Y.; Unpublished Daylight toolkit program Rotatable bonds: Single bonds not terminal, not in a ring, not secondary amide, not CX_3 (X = halogen); 2008.
17. Faber, K.; Kowalski, B. R. *J. Chemom.* **1997**, *11*, 181–238.
18. Topliss, J. G.; Edwards, R. P. *J. Med. Chem.* **1979**, *22*, 1238–44.
19. Golbraikh, A.; Tropsha, A. *J. Comput.-Aided Mol Des* **2002**, *16*, 357 69.
20. Clark, R. D.; Fox, P. C. *J. Comput.-Aided Mol. Des.* **2004**, *18*, 563–76.
21. SAMPLS. SAMple-distance Partial Least Squares; QCPE 650, version 1.3. Quantum Chemistry Program Exchange, Indiana University. Bloomington, IN. 1994.

8 Strategy for the Statistical Evaluation of a Data Set of Related Molecules

This chapter will illustrate the steps to follow in the regression or partial least-squares analysis of a data set. The discussion starts with preparing the data for the analysis, covers the straightforward initial stages of both linear and nonlinear methods, and finally moves on to the more complex tricks of the trade and pitfalls of the methods. There is no standard protocol for using these methods.[1–3]

One outcome of quantitative structure–activity relationship (QSAR) analyses is the suggestion as to which molecular properties are related to potency; in automated methods this exploration is called *variable selection*. The first phase of variable selection is accomplished once the user prepares the input data because often no additional properties will be considered. The subsequent statistical analyses determine the form of, and quantitative relationships (if any) between, these input properties and potency. Thus, variable selection is the cornerstone of all QSAR methods, even though it is more often discussed for methods that automatically examine large data sets.

In light of the ability of the various methods to identify those properties that are related to biological potency, it might be tempting to examine hundreds or thousands of potential predictor properties. Such an analysis runs the risk of chance correlations: examining so many predictors that a few happen by chance to be statistically significant. Topliss and Edwards examined this issue in great detail by generating sets of random numbers and examining the multiple regression equations that could be generated from them.[4,5] They found that, for example, for data sets of 10 or 20 observations, the probability of a statistically significant chance correlation rises from 30% if three predictors are examined to 80–90% if 15 are examined. In the case of 20 observations and 15 predictors, the maximum R^2 was 0.97 and the mean R^2 was 0.45. If 30 possible predictors are examined to explain 100 observations, the probability of a chance correlation is 99%, although the maximum R^2 falls to 0.52. These results suggest that it is advisable to decide in advance of the computational analysis how many and which properties will be used as possible predictors. Choosing a conservative number will support adding new properties to improve the fit.

Although there are many different computer programs that could be used for the analysis, the discussion will use an SAS statistical package[6] as an example. For multiple regression analyses with only a few predictors, Microsoft Excel[7] may be all that is required.

I. PREPARING THE DATA SET FOR ANALYSIS

The first suggestion is to include all of the molecules that are part of the data set with blanks left for missing molecular properties or potency. When a provisional model has been chosen, it may be possible to include molecules for which only irrelevant parameter values are missing.

For multiple regression, nonlinear regression, and PLS, the data set may contain as predictors either physical properties, indicator values for the presence or absence of particular groups, or a combination of the two. For CoMFA PLS, the data set contains the matrix of energies at various lattice points surrounding the molecules with the optional addition of physical properties or indicator variables.

Second, it is a good practice to carefully document any computational analysis. This documentation starts with the input data set. Each observation should have a clear label and, if possible, a second column of the SMILES[8] of the structure or of the variable substituents. Labels for the molecular properties should be explanatory and unambiguous. Additionally, the source and composition of each data set used for an analysis should be carefully documented because often an investigation involves more than one computational analysis. For example, analyses might examine alternative molecular descriptors such as 2D versus 3D examples; different subsets of a larger data set; different biological responses for the same set of molecules; or different sets of molecules that produce the same biological endpoint. Each of these data sets should carry a documentation that includes, at a minimum, the source of the data and the date that the set was prepared.

Third, once the data is in the computer but before the actual modeling, it is helpful to identify errors or outliers in the data and provide insights that will improve the quality of the statistical modeling. This would involve calculating simple statistics such as the standard deviation of each molecular property, examining the relationships between the properties by their correlations and principal components, and making various plots of the data. The results may help identify properties that are constant within the data set, identify errors in the data, and gain an impression of the relationships between potency and molecular properties. Such examinations are especially important when dealing with a data set of hundreds or more observations. Table 8.1 summarizes how these procedures highlight problems with the data set.

From Table 8.1 we can see that the procedures easily highlight outliers or errors in the data. If the data is not an error, then the apparent outlier should be considered for deletion. For example, Figure 8.1 illustrates a data set for which an outlier determines the statistical relationship. If the calculated log P and measured log $(1/C)$ values are correct, then the figure might suggest that either the molecule should be omitted from the analysis or the log $(1/C)$ of additional compounds of intermediate or high log P values should be measured.

TABLE 8.1

Symptoms of Problems for Regression or PLS Analysis of a Data Set

Problem:	There is a true outlier or an error in the data	A property does not vary	Some properties are not independent	It will be difficult to discover one quantitative model
Observation				
Value of the standard deviation of a property or log potency	It is very large	It is approximately equal to the error of the calculation	It provides no information	The standard deviation of the measured log potency is approximately equal to its measurement error
Plot of properties versus each other	One point is outside the range of the others	There is no variation in that property	Correlations are seen	Correlations are present—PLS may produce a model
Plot of properties versus potency	One point is outside the range of the others	No relationship is seen	There is a similar location of molecules on plots of different properties	Molecules are not uniformly distributed or there are nonlinear relationships. See Figures 8.1 and 8.2.
Values of the eigenvalues of the correlation matrix of the properties	Only indirect information	There are fewer significant eigenvalues than number of properties or observations	There are fewer significant eigenvalues than number of properties or observations	Correlations are present—PLS may produce a model
Plots of loadings of properties on principal components	Only indirect information	The variable does not contribute to any component	Certain properties are close in all the plots	Correlations are present—PLS may produce a model
Plots of scores of molecules on principal components	One point is outside the range of the others	No information	No information	Molecules are not uniformly distributed or nonlinear relationships

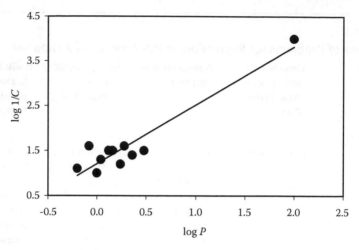

FIGURE 8.1   A plot that illustrates an influential observation at the top right of the figure.

Table 8.1 also shows procedures that identify properties that do not vary in the data set. Omitting such properties simplifies regression analysis because fewer combinations will need to be considered.

However, multiple measures can provide insight into the possibility that it may be possible to derive a quantitative model: A small standard deviation of the biological potency will limit the maximum R^2 of any model. Correlations between predictor properties complicate the understanding of regression results and hence suggest that partial least-squares analysis be used. Nonlinear relationships as seen in Figure 8.2 may dictate that a different form of the equation be fitted.

Lastly, before the actual modeling is performed it should be decided if there are enough molecules in the data set to divide it into training and test sets. As discussed

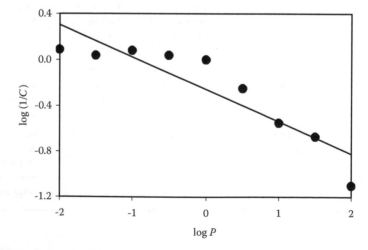

FIGURE 8.2   A plot that illustrates an apparently nonlinear relationship.

in Chapter 7, no matter how the test set is chosen, in no case should a molecule in the test set have identical molecular properties to one in the training set. Although certain analysis procedures include the automatic selection of training and test sets, sometimes called *validation sets*, it is advisable to retain an independent test set that is not used in any way during the analysis.

II. FINDING THE IMPORTANT MULTIPLE LINEAR REGRESSION EQUATIONS FOR A DATA SET

The equation that contains all of the predictors provides the highest R^2 and approximately the smallest s that can be achieved with a particular data set. If these statistics are not satisfactory, then the problem is not how to find the best equation, but rather what to do next. Usually this involves either considering other molecular properties, but any of the strategies listed in Section F might be used.

Because an analysis of a data set may involve fitting and comparing many statistical models, it is important to carefully document each model. Complete documentation of a model includes a listing of the input plus transformed data; the values of R^2, s, n, the coefficients of the predictors and their standard errors; the source of the data; and the date of the analysis.

A. THE PROBLEM OF MULTIPLE POSSIBLE EQUATIONS

QSAR models are developed both to suggest the properties of molecules that are responsible for biological activity and also to forecast the potency of as yet untested compounds. Hence, it is important that these statistical models be as robust and unambiguous as possible. The problem is not only to find the best equation (assuming that we could establish a criterion for best), but also to discover if there are other equations that use different molecular properties but that are not substantially different statistically from the best. For example, in a series of molecules that differ only in the length of an alkyl substituent, there will be a correlation between hydrophobicity and molar refractivity. As a result, there might be one equation that relates potency to the π hydrophobicity descriptor, but another of approximately equal R^2 that relates potency to the MR bulk descriptor and very different predictions of potency for analogs that have polar substituents. In other situations, a property may contribute to a model only in the presence of a second property: An example is the dependence of log $(1/C)$ in Equation 6.45 on both log P and (log $P)^2$. Because of the possibility that more than one equation may fit the data, it is convenient to use software that supports an efficient search through the possible equations.

Sometimes it is not clear if adding another term to a regression equation is appropriate. Table 8.2 lists a number of criteria that can be used to decide whether doing so is justified, even if the new term is statistically significant.[9–11] Computer programs for regression analysis will often provide one or more of these statistics.

Although it is desirable to examine all possible equations, this becomes impractical if there are more than a few descriptors. For a data set that contains p possible

TABLE 8.2

Statistical Criteria for the Addition of a Term to a Regression Equation
(n observations, p predictors)

Statistic	Symbol	Value	Comment
Akaike Information Criterion[10]	AIC	$n\log s_p^2 + 2p$	To include the pth variable, AIC should decrease from the best $p-1$ model.
Bayes Information Criterion[11]	BIC	$n\log s_p^2 + p\log n$	To include the pth variable, BIC should decrease from the best $p-1$ model.
Adjusted R^2	R_a^2	$1 - \left(\dfrac{n-1}{n-p}\right) \times \left(1 - R_p^2\right)$	To include the pth variable, R_a^2 should increase from the best $p-1$ model.
Predicted Residual Sum of Squares	PRESS	$SS_{error,p,loo}$	To include the pth variable, the sum of squared errors from leave-one-out cross-validation should decrease.
Mallow's C_p Statistic[9]	C_p	$\dfrac{SS_{error,p-1}}{s_p^2} + 2p - n$	A good model will have C_p approximately equal to p.

descriptors, there are $2^p - 1$ equations that contain at least one of these. The number of possible equations do not represent a problem with 3 descriptors for which there are 7 possible equations; but for 10 descriptors, not all of the 1023 possible equations would likely be of interest. Various approaches to streamlining the discovery of the useful equations will be discussed in this section.

B.   THE ALL-EQUATION APPROACH

In this method, the computer calculates every possible equation and provides a limited output for each. Because the output is examined manually, the practical upper limit of this approach is approximately 10 predictors. It is helpful if the program orders the output in some manner. In the SAS GLM procedure, for example, the R^2 of the one-predictor equations are listed first, in order of increasing R^2; then the R^2 of the two-predictor equations in order of increasing R^2, etc. An example of the output from this procedure is shown in Table 9.7 in the next chapter. The C-QSAR program provides the complete equation for each case sorted by s. The SAS procedure has options to specify that certain predictors are to be included in every equation, that the output contains only the best n equations for each number of predictors, or that only equations containing a specified number of predictors should be included.

Calculating all possible equations gives the most complete overview of the data. For example, R^2 usually levels off as more predictors are included, indicating how many predictors are needed for a model. If there are several equally significant equations, then the predictors that are involved may be correlated with each other.

C.   STEPWISE REGRESSION

Various stepwise regression procedures identify the best equations for a set of data. In backward elimination methods the calculation starts with the equation that contains terms for each of the molecular predictors. The algorithm then deletes the least significant term, recalculates the equation, and continues to delete terms until a user-defined stopping point is reached. In forward selection methods, the calculation starts with the equation that contains the single best predictor property. The algorithm then adds the next most significant predictor to the equation and continues until no more terms are statistically significant. In mixed stepwise methods, predictors may be either added or removed at any stage of the calculation. Stepwise procedures sometimes use cross-validation, particularly leave-one-out, as part of the model selection process.

When stepwise methods are used for a data set, the equations should be calculated by both backward elimination and forward selection methods. Backward elimination methods identify equations that describe a relationship with some molecular property X that appears in the equation as the quadratic terms $aX - bX^2$. If neither a nor b alone is statistically significant, a forward selection procedure would miss this relationship. On the other hand, forward selection identifies situations in which a few properties explain most of the variance in the data and yet several others may also be statistically significant.

D.   FOLLOW-UP TO STEPWISE REGRESSION

The appropriate follow-up to stepwise procedures is essential. If the forward and backward selection procedures arrive at the same equation, then it probably represents the most highly significant equation. Attention would then be placed on discovering equations of fewer terms that predict potency well. One strategy is to calculate all equations in which one, and perhaps two, predictors of this best equation have been deleted. Less complex equations need not be considered if the addition of the last predictor resulted in a substantial increase in R^2 and a decrease in s. In this case, the strategy may be to explore if the addition of a different predictor would have produced essentially the same R^2 and s. These goals might be accomplished by using an all-possible-regression method with the specification of certain predictors to be included in all equations and/or a lower and upper limit on the number of predictors to include in the equations.

The stepwise calculations sometimes suggest predictors that could be combined. For example, the data set may have contained π hydrophobic terms for different positions of substitution. If the stepwise equations indicate that the coefficients of these are similar, the sum of π could be used in future equations.

Stepwise calculations may identify one or more predictors that never enter an equation. If preliminary stepwise calculations can eliminate just two predictors from consideration, then the number of all possible equations will be cut to approximately one fourth; rejection of three predictors will cut it to approximately one eighth. Thus, the stepwise investigation of a data set might be a precursor to calculating all possible regressions, but with only relevant predictors.

E. OTHER ALGORITHMS FOR VARIABLE SELECTION

Genetic and evolutionary algorithms present another method to search for significant combinations of predictors of biological potency.[12–14] These methods start with a population of regression equations that include properties selected at random. Each equation is based on a different set of predictors. The presence or absence of a predictor in a particular model is encoded into a bit string for that model. Each bit represents one potential predictor. If the predictor is used for the model, it is encoded as 1.0; if it is not used, as 0.0. The fitness of each model is often the R^2 or q^2. The algorithms then delete many of the least fit models and use those remaining to randomly generate additional models. The new models are generated by mutation of one of the retained models (changing the value of the bit for a particular predictor) or crossover of two models (using the original bit strings up to some point, and then using the other one to complete the string). The resulting models are evaluated, some of the least fit are deleted, new models are generated, etc. The important variables are identified as those that appear most often in the best models generated.

Neural network searches are also used for variable selection.[15–17] A neural network contains an input layer of nodes, each of which corresponds to one predictor. The hidden layers describe the relationships between these predictors and potency. The network is trained to optimize these relationships, essentially to select predictors that are important predictors of potency. Although the network itself can be used as the model, the network can also be used to identify the important descriptors for a stepwise or all-regression analysis.

A PLS calculation on the data will identify potential predictors that do not contribute to (load on) any of the latent variables. These predictors can be omitted from any further multiple regression analyses.

F. IMPROVING A REGRESSION EQUATION

Several strategies can help improve a regression equation that has a high standard deviation or rather low R. The problems might be that not every molecule can be included in one equation or that an additional property needs to be considered.

Residual plots are often helpful in investigating such problems.[2] For example, a plot of the residual versus a new predictor that will add significantly to a regression equation will show a positive or negative relationship. For a sample data set, adding CLOGP (Figure 8.3) to the equation might improve the fit, but CMR (Figure 8.4) probably will not. However, Figure 8.3 also suggests that there might be two subsets of observations.

In addition, plots of the residual versus either properties included in the model or ones being considered for inclusion may identify an outlier. Such an outlier will decrease the significance of the relationship that would fit most of the other

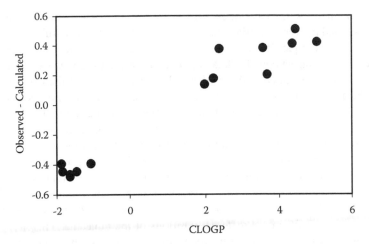

FIGURE 8.3   A plot that suggests that adding CLOGP to the equation will improve the fit of the data.

molecules. Then, of course, the problem is whether to omit the outlier and generate a better fitting equation or to try to understand why the molecule is an outlier.

A residual plot might show that there are two subsets in the data. For example, it might be observed that all *meta*-substituted molecules are less potent than predicted, and all *para*-substituted are more potent. Adding an indicator variable to distinguish between the two subsets would then result in a more significant equation.

Yet another plot, that of the residual versus log P or pK_a, may show curvature, suggesting that a model-based equation and nonlinear regression analysis might improve the fit.

A related technique for the analysis of resistant data is to subdivide the molecules into subsets and perform the statistical analysis on them separately. Such separations

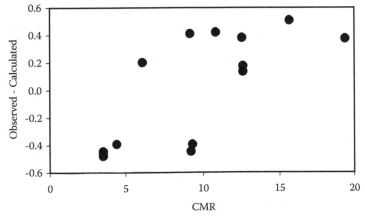

FIGURE 8.4   A plot that suggests that adding MR to the equation might not improve the fit of the data.

are logically made on the basis of chemical properties: alkyl- versus aryl-substituted analogs, secondary versus tertiary amines, or *meta-* versus *para-*substituted phenyl molecules, for example. If there are satisfactory regression equations for such subsets, they might suggest how the data could be described to produce an overall equation. Although this would usually involve adding one or more predictors, it would not risk artificial correlations[4,5] to the extent of searching through a number of possible predictors. Alternatively, the individual equations might suggest that the two cannot be combined. Predictions of new molecules would then be based on a smaller number of observations, but these predictions might prove to be valid.

Sometimes, a more satisfactory regression can be obtained by using different molecular properties. For example, the most common relationship between hydrophobicity and potency is the linear or nonlinear correlation of potency with the overall log P of the molecules. However, the relationship between potency and hydrophobicity might be different for substituents at different positions of the molecule. Therefore, if there are enough molecules in the data set, worthwhile information about the series may be discovered by considering the π-values of each variable position separately. However, a complication in deciding positions of substitution sometimes arises. For example, for a molecule in which the 2- and the 6-substituents of a phenyl ring are not identical, which π-values should be used for position 2 and for position 6? There is no set answer to this problem, although one solution is to consider the more hydrophobic substituent to be at position 2. This would correspond to hydrophobic binding to the target of only a part of the molecule. Sometimes, a clearer choice is highlighted if a list is prepared in which the molecules are listed in order of decreasing potency.

G. Indicator Variables

Indicator variables may be used to distinguish nonstructural features in a data set. If the same compounds had been tested in several biological assays, they may be used to examine the similarities of the various structure–activity relationships. To do so, the usual matrix of potency values and physical properties would be prepared, using one row for each molecule in each biological assay. Indicator variables to distinguish the various assays, one fewer than the total number, would be used to account for the fact that the potency had been measured in different assays. A comparison of the s of the total equation with those derived from data on only one assay system will show if the structure–activity relationships are the same for all tests. If so, the coefficients of the indicator variables show the relative sensitivity of the assays to the series.

In a similar manner, indicator variables may be used to combine the data on similar tests conducted by different investigators or under different conditions of pH, temperature, etc.

When data sets are combined and an indicator variable used, it is important that the structure–activity relationships of the total data set agree with those found on the individual sets. If the magnitude of the dependent variable for one set is different from that of the other, there may be a rather high R^2 and F for the relationship between potency and the indicator variable. This is merely an ANOVA problem in disguise. The important statistic to consider is s, which reflects the precision of estimates based on the equation.

As an example of this sort of use of indicator variables, we considered the several different biological properties of barbiturates. Hansch and Anderson reported the following equations:[18]

50% inhibition of *Arabica* cell division

$$\log (1/C_i) = 0.80 \log P_i + 1.08$$

$$R^2 = 0.91, s = 0.17, n = 19$$

(8.1)

50% inhibition of rat brain oxygen consumption

$$\log (1/C_i) = 1.04 \log P_i + 0.96$$

$$R^2 = 0.91, s = 0.20, n = 10$$

(8.2)

50% inhibition of NADH oxidation

$$\log (1/C_i) = 1.11 \log P_i + 1.24$$

$$R^2 = 0.85, s = 0.26, n = 6$$

(8.3)

Hypnotic activity

$$\log (1/C_i) = 2.09 \log P_i - 0.63 \log P_i^2 + 1.92$$

$$R^2 = 0.98, s = 0.14, n = 11$$

(8.4)

We combined all the data and used the indicator variable A to represent activity versus *Arabica*, B versus brain, and N versus NADH. That is, all three in vitro tests were compared with the rat hypnotic activity. The equation is

$$\log (1/C_i) = 0.70 \log P_i - 1.07 A - 0.83 B - 0.50 N + 2.33$$

$$R^2 = 0.76, s = 0.33, n = 46$$

(8.5)

From this equation we may conclude that the structure–activity relationships are more or less identical with the four tests. However, absolute potency is greatest in the test for hypnotic activity because the coefficients of the indicator values for the other tests are negative. The larger value of s from Equation 8.5 indicates that the data sets are not strictly additive. This might be due to the inclusion of analogs with supermaximal log P's in the rat test.

The advantage of such a combination of data is that it provides an overview of structure–activity relationships for more cases.

III.   USING NONLINEAR REGRESSION ANALYSIS

A.   Comparison of Linear and Nonlinear Regression

Chapter 6 made clear that fits of data to model-based structure–activity equations usually require nonlinear regression analysis. As noted in Chapter 7, nonlinear regression differs from linear in that the equations to are a complex function of the coefficients and hence cannot be solved explicitly; rather, an iterative procedure must be used.

B. SOME SPECIAL PROBLEMS WITH NONLINEAR REGRESSION ANALYSIS

Nonlinear methods may not be necessary for the data. As noted above, before any fits are considered, plots of each property as a function of potency should be inspected. In the case of plots with log P, the points should be identified by fraction un-ionized. If there are no obvious nonlinearities, the usual linear regression analysis strategy can be followed. Only if the original plots show obvious curvature, if the residuals from the linear fit show a curved trend with a physical property, or if there is a complex relationship between potency and degree of ionization should nonlinear fits be tried. If the compounds do not vary in degrees of ionization and if the upward and downward slopes of a nonlinear relationship are approximately equal, then fits to the empirical parabolic function in Equation 6.45 are also appropriate.

Input to nonlinear regression analysis includes the form of the equation and an initial estimate of the values of the coefficients. Sometimes, these estimates are derived from experience; for example, the long tradition of QSAR usually finds the coefficient of the relationship between log $(1/C)$ and log P to be somewhere between 0.5 and 1.0. Accordingly, our initial estimate of b for fitting Equation 6.9 might be 0.75. If preliminary plots suggest that the predominant determinant of potency is a nonlinear function of some property, then several initial guesses of the coefficients may be substituted into the function, and the data and predictions plotted to show which best fits the data. This is easily done in Microsoft Excel. A plot is helpful because the fit in each region of space can be examined and varied as the parameter of the equation is changed.

The preliminary plots should suggest which structure–activity equation seems to be most likely to be appropriate. Chapter 6 discussed this point in more detail. As in linear regression analysis, it is essential to compare fits to several alternative equations.

The equations in Chapter 6 or any derived may not be of the correct form to be fit. For example, if the molecules are ionizable, the appropriate function of pH and $\text{p}K_a$ is substituted for α. For example, if there were two aqueous compartments ($n = 2$) and one nonaqueous compartment ($m = 1$) for a cationic acid (amine), Equation 6.34 would become

$$\log(1/C_i) = \log \frac{1+zR_i}{1+zR_i+dP_i^c+\dfrac{1}{P^b a_1 u_{1,i}}+\dfrac{1}{P^b a_2 u_{2,i}}} + X \qquad (8.6)$$

R is the ratio of the fraction of molecule i ionized, $\alpha_{2,i}$, to that un-ionized, $1 - \alpha_{2,i}$ in the aqueous compartment (2) that contains the target biomolecule. This ratio is calculated from Equations 4.8 and 4.9:

$$R_i = \frac{\alpha_{2,i}}{1-\alpha_{2,i}} = \frac{10^{\text{p}K_{a,i}-\text{pH}_2}+1}{10^{\text{pH}_2-\text{p}K_{a,i}}+1} \qquad (8.7)$$

The variables u_1 and u_2 are the fraction un-ionized in aqueous compartments 1 and 2. They are also evaluated from Equation 4.9:

$$u_{1,i} = \frac{1}{10^{pK_{a,i}-pH_1}+1} \tag{8.8}$$

$$u_{2,i} = \frac{1}{10^{pK_{a,i}-pH_2}+1} \tag{8.9}$$

Hence, the equation to be fit is

$$\log(1/C_i) =$$

$$\log \frac{1 + z\dfrac{10^{pK_{a,i}-pH_2}+1}{10^{pH_2-pK_{a,i}}+1}}{1 + z\dfrac{10^{pK_{a,i}-pH_2}+1}{10^{pH_2-pK_{a,i}}+1} + dP_i^c + \dfrac{1}{P_i^b a_1 \dfrac{1}{10^{pK_{a,i}-pH_1}+1}} + \dfrac{1}{P_i^b a_2 \dfrac{1}{10^{pK_{a,i}-pH_2}+1}}} + X \tag{8.10}$$

The dependent variable is $\log(1/C_i)$; the independent variables are $pK_{a,i}$, pH_1, and P_i; and the coefficients to be fit are z, pH_2, d, c, a_1, a_2, b, and X.

However, it is possible that the values of all eight coefficients are not independent. If they are not, the matrix for solving the nonlinear equation will be singular, and it will not be possible to provide unique answers for each coefficient. This situation arises when some of the terms in an equation dominate the relationship. For example, in Equation 8.10 if $z = 0$ and the term $pK_a - pH_2 > 1$, then Equation 8.10 reduces to Equation 8.11:

$$\log(1/C_i) = \log \frac{1}{1 + dP_i^c + \dfrac{1}{P_i^b a_1 \dfrac{1}{10^{pK_{a,i}-pH_1}+1}} + \dfrac{1}{P_i^b a_2 \dfrac{1}{10^{pK_{a,i}-pH_2}}}} + X$$

$$= \log \frac{1}{1 + dP_i^c + \dfrac{1}{P_i^b a_1 \dfrac{1}{10^{pK_{a,i}-pH_1}+1}} + \dfrac{1}{P_i^b a_2 \dfrac{10^{pH_2}}{10^{pK_{a,i}}}}} + X \tag{8.11}$$

Thus, a_2 and pH_2 cannot be independently fitted because they occur only in combination. To prevent the selecting of correlated coefficients, preliminary estimates would be used to calculate the expected values of the individual terms in the equation. If some of these values are much smaller than others, then the reduced equation would be investigated.

On the other hand, the projected values of the terms in an equation may suggest that the shape observed in the plots cannot be achieved with the selected function.

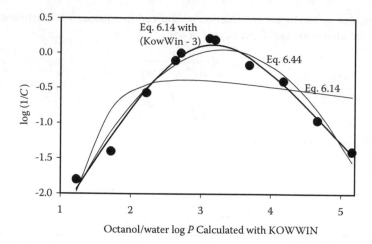

Octanol/water log P Calculated with KOWWIN

FIGURE 8.5 A plot showing the fit of the inhibition of dopamine β-hydroxylase by fusaric acid derivatives. The pK_a within the series is constant. Equation 6.44 describes a parabolic relationship between potency and log P. The best fit of the data, the heavy line, is found using Equation 6.14 if the fit is based on log P − 3 rather than log P. This equation is based on Scheme 6.2, which contains a nonaqueous, an aqueous, and a receptor compartment. This example will be discussed further in Chapter 10.

For example, Equation 6.14 is most easily fitted if the apparent optimum log P is subtracted from the individual log P_i values. Figure 8.5 shows an example of the results of this strategy.

Because nonlinear regression analysis identifies the best estimates of the coefficients by an iterative procedure, the convergence criterion and the maximum number of iterations that should be tried must be specified. It is assumed that the minimum SS_{error} has been reached (the estimates "converged") if a new iteration does not decrease SS_{error} by more than the convergence criterion, typically 0.1%. The maximum number of iterations can be set as high as 1000 because usually the function converges in fewer iterations.

There might be predictors that enter the equation only in a linear fashion, for example, E_s of Equations 6.35 and 6.36. These can be identified by plots of the residuals of the nonlinear fit versus the candidate predictors, or by stepwise or all-possible regression analysis using the residual as the dependent variable. The coefficient of any significant predictor would be used as an initial estimate in the nonlinear regression to reoptimize the whole function.

IV. FITTING PARTIAL LEAST-SQUARES RELATIONSHIPS

As discussed in Chapter 7, a PLS analysis investigates the relationship between a property to be predicted, potency for example, and a large number of predictor properties. Because no variable selection is involved, PLS analyses are initially more straightforward than ordinary or nonlinear least regression analyses. One must decide only how many latent variables to include in the model and which predictor properties to consider.

Just as with regression analysis, as more and more latent variables are included in the PLS fit, the R^2 of the model continually increases. For this reason, R^2 cannot be used to decide how many latent variables are important. Instead, the usual procedure is to base the decision on the results of leave-one-out cross-validation—for example, to choose the number of latent variables that produce the highest cross-validated q^2. However, we also require that adding a particular latent variable decrease the cross-validated standard error by at least 5%.

Because PLS can discover the important relationships, often no variable selection is used. On the other hand, because PLS includes the contribution of every predictor, it is susceptible to noise or irrelevant predictors.[19]

Omitting one or more properties can speed up a PLS calculation, particularly a leave many out cross-validation. Any property that does not vary can be omitted: This criterion is used to eliminate columns for a CoMFA analysis by dropping those columns that record constant interaction energies of all the molecules with the probe at a certain lattice point. Such a constant energy could arise either because the point is inside the surface of every molecule and so the steric repulsion energy is always at the truncation level, or because it is far away from every molecule and so both the steric and electrostatic interactions are zero.

If the analysis includes physical properties as predictors of potency, the plots suggested in Section I might suggest that certain properties may not be important to include. After a provisional PLS model is derived, the loadings of the predictors on the latent variables may suggest other predictors to omit from a later model.

Predictor selection by the same algorithms as are used in regression analysis has also been reported.[13,20–22] In contrast, the program GOLPE uses statistical design criteria to optimize the search for relevant predictors.[23]

V.   TESTING THE VALIDITY OF A COMPUTATIONAL MODEL

The statistical significance of each term in a model is a basic prerequisite for a valid model. In addition, Chapter 7 detailed three tests of the reliability of a computational model: (1) the ability of the model to predict the potency of molecules from a test set, (2) the closeness of q^2 to R^2 for the model, and (3) the difference between the R^2 from the model and that from models based on y-scrambled data. In addition, the criteria listed in Table 8.2 might be helpful. Once all the above criteria are met, further analysis is needed to gain confidence that the model will be useful.

The first test is to be sure that the model has not overfitted the data. This occurs if the $s_{replicate}^2$ value for the biological test is larger than the s^2 of the provisional model. Overfitting means that some of the experimental error has been explained by the equation.

Another important test is that the provisional model should predict the inactivity of inactive molecules or the potency of molecules previously omitted for apparently good reason. Such predictions, whether accurate or not, give a measure of the scope of the equation.

A model might be less valid than the statistics suggest if R^2 increases by only a small amount with the addition of some property or latent variable. Plots of this property or latent variable versus the residuals from the model without the term in question might show that the additional term is significant primarily because it more precisely predicts the potency of a single molecule. In such a case, that molecule could be omitted to see if models based on the remaining molecules result in the same coefficients and R^2 or s as the original (with and without the predictor under suspicion). Alternatively, additional analogs might be synthesized and tested to assess the relevance of the property in question.

It can be difficult to decide if a particular molecule should be omitted from a model or not. The problem is that there is one easy way to get an equation of high R^2 and low s from a set of data. To do so, the best fit of the potency to the properties of interest is first discovered. Then the residual for each molecule is calculated, all molecules that do not fit this relationship within the predetermined value of s are omitted, and the fit for the revised data set is calculated. By such a procedure, any relationship can be "proved." Thus, a careful worker will state which molecules of the original data set are not included in the final equations, the basis on which they were omitted, and their predicted potency. At best, the omission of molecules is a negative comment on the quality of the biological data or the usefulness of the structure–activity considerations; at worst, such an omission has merely created an artifact that satisfies its creator but has no power of prediction.

VI. LESSONS LEARNED

There are straightforward techniques to help one identify important regression or PLS models of a data set. Key to discovering useful models is the careful preparation of the structures and molecular properties that are to be used. Equally important is the thorough examination of the stability of any relationship by cross-validation, training versus test sets, and scrambling the biological potency values.

REFERENCES

1. Draper, N. R.; Smith, H. *Applied Regression Analysis*, 3rd ed. Wiley: New York, 1998.
2. Faraway, J. J., *Practical Regression and Anova Using R;* 2002. http://lib.stat.cme.edu/R/CRAN (accessed Feb. 6, 2008).
3. Eriksson, L.; Johansson, E.; Kettaneh-Wold, N.; Trygg, J.; Wikström, C.; Wold, S. *Multi- and Megavariate Data Analysis. Part I: Basic Principles and Applications*, 2nd ed. Umetrics AB: Umeå, 2007.
4. Topliss, J. G.; Edwards, R. P. In *Computer-Assisted Drug Design*; Olson, E. C., Christoffersen, R. E., Eds. American Chemical Society: Washington, DC, 1979, pp. 131–45.
5. Topliss, J. G.; Edwards, R. P. *J. Med. Chem.* **1979**, *22*, 1238–44.
6. SAS/STAT, SAS Institute. Cary NC. http://www.sas.com/technologies/analytics/statistics/stat/.
7. Microsoft® Excel 2000, Microsoft. Redmond, WA. 2000.
8. http://www.daylight.com/smiles/index.html; 2007. (accessed Nov. 25, 2007). Daylight Chemical Information Systems, Inc.
9. Mallows, C. L. *Technometrics* **1973**, *15*, 661–75.
10. Akaike, H. *IEEE Transactions on Automatic Control.* **1974**, *19*, 716–23.
11. Schwarz, G. *Ann. Stat.* **1978**, *6*, 461–64.

12. Clark, D. E.; Westhead, D. R. *J. Comput.-Aid. Mol. Des.* **1996**, *10*, 337–58.
13. Kubinyi, H. *J. Chemom.* **1996**, *10*, 119–33.
14. Broadhurst, D.; Goodacre, R.; Jones, A.; Rowland, J. J.; Kell, D. B. *Anal. Chim. Acta* **1997**, *348*, 1–3.
15. Wikel, J. H.; Dow, E. R. *Bioorg. Med. Chem. Lett.* **1993**, *3*, 645–51.
16. Tetko, I. V.; Villa, A.; Livingstone, D. J. *J. Chem. Inf. Comput. Sci.* **1996**, *36*, 794–803.
17. Livingstone, D. J.; Manallack, D. T.; Tetko, I. V. *J. Comput.-Aid. Mol. Des.* **1997**, *11*, 135–42.
18. Hansch, C.; Anderson, S. M. *J. Med. Chem.* **1967**, *10*, 745–53.
19. Clark, M.; Cramer III, R. D. *Quant. Struct.-Act. Relat.* **1993**, *12*, 137–45.
20. Lindgren, F.; Geladi, P.; Rannar, S.; Wold, S. *J. Chemom.* **1994**, *8*, 349–63.
21. Hasegawa, K.; Miyashita, Y.; Funatsu, K. *J. Chem. Inf. Comput. Sci.* **1997**, *37*, 306–10.
22. Hasegawa, K.; Funatsu, K. *SAR QSAR Environ. Res.* **2000**, *11*, 189–209.
23. Cruciani, B.; Clementi, S.; Baroni, M. In *3D QSAR in Drug Design: Theory Methods and Applications*; Kubinyi, H., Ed. ESCOM: Leiden, 1993, pp. 551–64.

9 Detailed Examples of QSAR Calculations on Erythromycin Esters

As an example of how one chooses the relevant regression equations, this chapter will describe in detail the strategy that led to the published quantitative structure–activity relationship (QSAR) models of esters of erythromycin.[1] Table 1.1 and Structures 1.9 and 1.10 list the structures and biological activity of the compounds considered in this chapter. Chapter 10 will continue by summarizing prior and subsequent analyses, the predictivity of the various QSARs, and conclusions from the various approaches.

I. ANTIBACTERIAL POTENCIES VERSUS *STAPHYLOCOCCUS AUREUS*

The antibacterial potency of the analogs had been measured in a whole-cell turbidometric assay, which follows the growth of bacteria in the presence and absence of an antibiotic. The raw potency was reported on a milligram basis relative to erythromycin A, which was assigned a potency of 1000. For the QSAR analysis, we corrected the relative potencies for the differences in molecular weight and used a logarithm of the corrected potency for the variable log $(1/C_{SA})$.

II. CALCULATION OF THE MOLECULAR PROPERTIES

A. PARTITION COEFFICIENTS

We expected that there would be a correlation with the log P's of the molecules because (1) the relative potency had been measured in a whole-cell assay in which the molecules had to penetrate the cell wall and cell membranes to reach the target sites, and (2) log P is often correlated with cell penetration.[2] In order to be able to calculate the log P's (before computer programs to do so became available), we needed to measure partition coefficients to establish π values for the following changes: (1) addition of a hydroxyl group at position 12 (erythromycin A versus erythromycin B analogs), (2) change of an -OH to -OC(=O)H, and (3) the addition of a methylene group to an ester.

A problem for the log P determination was that the compounds are not soluble and stable at the pH values at which the amino group is not protonated, which are the typical pHs at which log P would be measured. Because all analogs except the

175

TABLE 9.1
Measured Octanol–Water Partition Coefficients

Compound	Structure	Log P^a
1.09	Erythromycin A	2.54
1.10	Erythromycin B	3.07
1.14	4″-Formylerythromycin A	2.64
1.19	4″-Formylerythromycin B	3.05
1.23	11-Acetylerythromycin B	3.30
1.25	4″-Acetylerythromycin A	2.85
1.29	4″-Acetylerythromycin B	3.23

[a] Adjusted values as explained in the text.
Source: Martin, Y. C. et al. *Med. Chem.* **1972**, *15*, 635–38.

2′-esters would have identical pK_a values, we decided to measure log P's of erythromycins A and B and several 4″ or 11-esters at pH 8.0, assuming that only the neutral form partitions into the octanol and adjusting the observed log P by using Equation 4.9 and a pK_a value of 8.6. To quantitate the concentration of compound, we used a colorimetric method.[1] The measured partition coefficients are shown in Table 9.1.

Because the log P values of the 4″-formyl esters are essentially equal to the corresponding nonesterified analog, we wondered if the ester had hydrolyzed during the log P determination. Therefore, we repeated the partitioning to check the aqueous

TABLE 9.2
π Values Calculated from the Measured Log P Values

Structural Change	Analog	Parent	π
-OH to -OC(=O)H	1.14	1.09	+0.10
	1.19	1.10	−0.02
		Mean	+0.04
-OH to -OC(=O)CH$_3$	1.25	1.09	0.31
	1.29	1.10	0.16
	1.23	1.10	0.23
		Mean	0.23 ± 0.08
-OC(=O)H to -OC(=O)CH$_3$	1.25	1.14	0.21
	1.29	1.19	0.18
	1.23	1.19	0.25
		Mean	0.21 ± 0.03
Erythromycin A to B	1.10	1.09	0.53
	1.29	1.25	0.38
	1.19	1.14	0.41
		Mean	0.44 ± 0.08

Source: Martin, Y. C. et al. *J. Med. Chem.* **1972**, *15*, 635–38.

TABLE 9.3
π Values Used to Calculate Log P

Structural Change	π
12-H to 12-OH	−0.44
-OH to −OC(=O)CH$_3$	0.00
-OH to −OC(=O)CH$_3$	0.23
-OC(=O)H to −OC(=O)CH$_3$	0.23
−OC(=O)CH$_3$ to −OC(=O)CH$_2$CH$_3$	0.23

phase for the presence of hydrolyzed compound. The analysis was performed by thin-layer chromatography with the spots visualized by the presence of antibacterial activity. The compounds had not hydrolyzed.

Table 9.2 shows the reproducibility of the change in log P with change in structure. From these Δ log P values, we calculated the π values shown in Table 9.3. These values and the log P of erythromycins A and B were used to calculate the log P values of all analogs.

The π values measured for the erythromycin analogs are attenuated compared to those reported for simpler molecules. For example, although the π value for a typical aliphatic -OH group is −1.12,[3] that of the 12-OH of erythromycin A is −0.44 (the difference between the log P of erythromycin A and B). Presumably the value is less because the 12-OH can form a hydrogen bond to the 11-OH group.[4] This is shown in Figure 9.1, which displays the hydrogen bonds in the x-ray structure of 6-O-methylerythromycin.[5] On the other hand, the π value for adding a methyl group to the ester side chain is 0.23 compared to the usual π of 0.50.[3] It has been postulated that these alkyl groups form a hydrophobic interaction with the hydrophobic face of the molecule, visible in Figure 9.2.[4]

The unique nature of the environments of the variable substituents in these analogs is also highlighted by comparing the log P values calculated from the experimental measurements with those calculated by CLOGP[6] (Figure 9.3). Although the

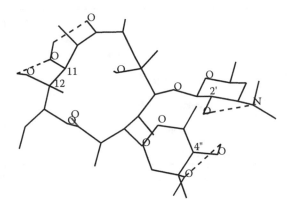

FIGURE 9.1   A view of the x-ray structure of 6-O-methylerythromycin that shows some of the intramolecular hydrogen bonds.

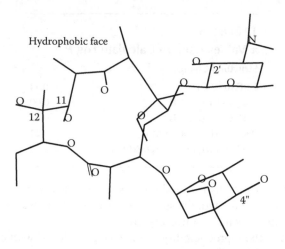

FIGURE 9.2   A view of the x-ray structure of 6-O-methylerythromycin that shows the possible hydrophobic interactions in the macrocyclic macrolide ring

two sets of log P values are reasonably correlated, note that the slope of the correlation is not unity: Whereas the CLOGP values varies from 1.8 to 5.5 for a span of 3.5 log units, those based on measured values vary only from 2.6 to 3.8 for a span of 1.2 log units. Indeed, values calculated with CLOGP, and those calculated from the experimental measurements are correlated with an R^2 of 0.82 and a standard error of 0.42. The corresponding plot for KowWin[7] (Figure 9.4) shows more scatter. Neither method captures the structure-hydrophobicity trends of the measured log P values.

B.   OTHER 2D PROPERTIES FOR TRADITIONAL QSAR

There are only two different substituents at position 12 (-H or –OH, erythromycin B versus A); hence, it is not possible to decide if any difference in potency is a result

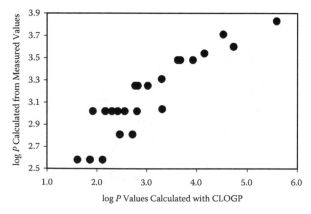

FIGURE 9.3   The relationship between the log P calculated from measured values and that calculated from the CLOGP program.

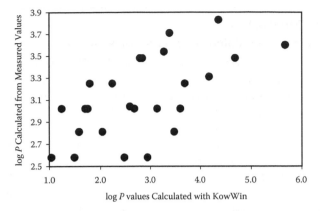

FIGURE 9.4 The relationship between the log P calculated from measured values and that calculated from the KowWin program.

of changes in hydrophobic, electronic, or steric properties. As a result, this structural difference is properly indicated in a regression equation by an indicator variable. In Table 9.4 the property A is 1.0 for erythromycin A derivatives and 0.0 for erythromycin B derivatives.

Aside from changes in hydrophobicity, an important consequence of esterification is the change in the direction and strength of potential hydrogen bonds. Thus, we used indicator variables to specify the presence or absence of esterification of the 4″ and 11 hydroxyl groups: D4 and D11 are set equal to 1.0 if the position is esterified and 0.0 if it is not. The properties used in the calculation are listed in Table 9.4.

TABLE 9.4
Physical Properties and Potencies of Erythromycin Esters

Structure	Name	Log P	D4	D11	A	Log $(1/C_{SA})$
1.09	Erythromycin A	2.58	0	0	1	3.00
1.11	2′-Formylerythromycin A	2.58	0	0	1	2.91
1.10	Erythromycin B	3.02	0	0	0	2.82
1.12	2′-Formylerythromycin B	3.02	0	0	0	2.78
1.13	2′-Acetylerythromycin A	2.81	0	0	1	2.75
1.14	4″-Formylerythromycin A	2.58	1	0	1	2.72
1.15	2′,4″-Diformylerythromycin A	2.58	1	0	1	2.71
1.16	11-Formylerythromycin B	3.02	0	1	0	2.71
1.17	2′-Acetyl-4″-Formylerythromycin A	2.81	1	0	1	2.70
1.18	2′-Acetylerythromycin B	3.25	0	0	0	2.54
1.19	4″-Formylerythromycin B	3.02	0	0	0	2.54
1.20	2′,4″-Diformylerythromycin B	3.02	1	0	0	2.50
1.21	4″,11-Diformylerythromycin B	3.02	1	1	0	2.42
1.22	2′,4″,11-Triformylerythromycin B	3.02	1	1	0	2.39

(Continued)

TABLE 9.4 (CONTINUED)
Physical Properties and Potencies of Erythromycin Esters

Structure	Name	Log P	D4	D11	A	Log (1/C_SA)
1.23	11-Acetylerythromycin B	3.25	0	1	0	2.29
1.24	2'-Acetyl-4"-Formylerythromycin B	3.25	1	0	0	2.28
1.25	4"-Acetylerythromycin A	2.81	1	0	1	2.24
1.26	2',4"-Diacetylerythromycin A	3.04	1	0	1	2.15
1.27	2',11-Diacetylerythromycin B	3.48	0	1	0	2.11
1.28	2',4"-Diacetylerythromycin B	3.48	1	0	0	2.04
1.29	4"-Acetylerythromycin B	3.25	1	0	0	2.02
1.30	4"-Propionylerythromycin B	3.31	1	0	0	2.02
1.31	2'-Acetyl-4"-propionylerythromycin B	3.54	1	0	0	1.87
1.32	2',11-Diacetyl-4"-formylerythromycin B	3.48	1	1	0	1.65
1.33	4",11-Diacetylerythromycin B	3.48	1	1	0	1.51
1.34	2',4",11-Triacetylerythromycin B	3.71	1	1	0	1.39
1.35	4",11-Dipropionylerythromycin B	3.60	1	1	0	1.38
1.36	2',4",11-Tripropionylerythromycin B	3.83	1	1	0	1.18

III. STATISTICAL ANALYSIS FOR TRADITIONAL QSAR

A. SIMPLE TWO-VARIABLE RELATIONSHIPS

As discussed in Chapter 8, plots of the data can reveal problems that can prevent a successful QSAR analysis. They can highlight outlier observations, suggest nonlinear relationships, and identify properties that do not appear to be related to potency.

The relationship between potency (log $(1/C_{SA})$) and each of the physical properties is plotted in Figures 9.5–9.9. Each plot supports a possible relationship, and none show outliers. Nonlinearity is not seen in the relationship of potency with log P (Figure 9.5). The correlation of each of the physical properties with log $(1/C_{SA})$ is listed as the top line of Table 9.5: indeed, all correlations are statistically significant.

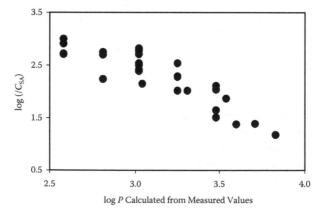

FIGURE 9.5 The potency of the erythromycin esters as a function of log P.

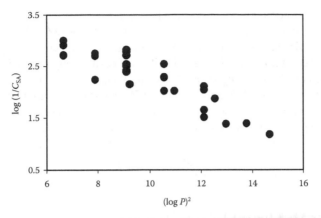

FIGURE 9.6   The potency of the erythromycin esters as a function of $(\log P)^2$.

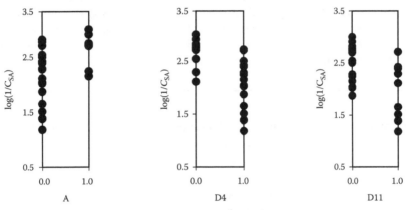

FIGURE 9.7   The potency of the erythromycin esters as a function of A.

FIGURE 9.8   The potency of the erythromycin esters as a function of D4.

FIGURE 9.9   The potency of the erythromycin esters as a function of D11.

TABLE 9.5
Correlations between Properties of Erythromycin Esters

	Log P	$(\text{Log } P)^2$	A	D4	D11
Log $(1/C_{SA})$	−0.89***	−0.90***	0.48*	−0.56*	−0.56*
Log P	1.00	1.00***	−0.75***	0.29	0.54*
$(\text{Log } P)^2$	1.00***	1.00	−0.73***	0.30	0.54*
A	−0.75***	−0.73***	1.00	−0.02	−0.47
D4	0.29	0.30	−0.02	1.00	0.09
D11	0.54*	0.54*	−0.47	0.09	1.00

*The correlation is significant at a probability <0.01.
***The correlation is significant at a probability <0.0001.

Table 9.5 also contains the correlation between each pair of molecular properties. Note that log P and $(\log P)^2$ are correlated with an R-value of 1.00, which indicates that they each contain the same information. As a result, we expect that only either log P or $(\log P)^2$, and not both, will be statistically significant in any regression equation.

The correlation between log P and A, although highly statistically significant, is not of such concern because R^2 is only $-0.75^2 = 0.56$, confirming that log P and A include independent types of information with respect to physical properties. The other correlations between physical properties are lower and therefore even less cause for concern.

The correlation between log P and $(\log P)^2$ and that between log P and A was remedied when less hydrophobic analogs were synthesized. The analysis of this larger set of analogs will be discussed in Chapter 10.

B.   STATISTICS FOR THE ALL-VARIABLE EQUATION

Table 9.6 lists the output from the Microsoft Excel[8] calculation of the equation that uses all properties to predict log $(1/C_{SA})$. The high R^2, low standard error of prediction, and details of the analysis of variance suggest a highly significant fit. However, the detailed statistics for the regression equation show that neither log P nor $(\log P)^2$ achieve the P-value of 0.05 required for statistical significance, confirming that log P and $(\log P)^2$ cannot be included in the same regression equation.

TABLE 9.6

The Statistics for the Equation that Includes All Predictors of Log $(1/C_{SA})$

Regression Statistics

Multiple R	0.97
R^2	0.95
Adjusted R^2	0.94
Standard error	0.13
Observations	28

ANOVA

	Degrees of Freedom (DF)	SS	MS	F-value	Significance F
Regression	5	6.48	1.30	79.48	0.00
Residual	22	0.36	0.02		
Total	27	6.84			

	Regression Equation				95% Confidence Limits	
	Coefficients	Standard Error	t Statistic	P-Value	Lower	Upper
Intercept	4.54	2.53	1.79	0.09	−0.71	9.80
Log P	0.09	1.55	0.06	0.96	−3.13	3.30
$(\log P)^2$	−0.22	0.24	−0.94	0.36	−0.72	0.27
D4	−0.29	0.06	−5.24	0.00	−0.40	−0.17
D11	−0.16	0.06	−2.57	0.02	−0.29	−0.03
A	−0.30	0.11	−2.77	0.01	−0.52	−0.07

C.   IDENTIFICATION OF THE BEST EQUATIONS

The next step was to use the SAS[9] RSQUARE option of the GLM procedure to summarize all possible equations. The results are in Table 9.7. The best single-variable predictors of potency are log P and (log $P)^2$. In addition, D4 and A appear to improve the equation, and D11 might also be significant. As expected, log P and (log $P)^2$ in the same equation is no better than either alone.

These results aided the selection of the QSAR equations that are summarized in Table 9.8. The equations were selected to include log P rather than (log $P)^2$ and to add terms only if doing so increased R^2 over the best equation with one fewer term.

TABLE 9.7
R^2 for All Possible Regression Models for Log (1/C_{SA}) ($N = 28$)

Number in Model	R^2	Variables Included
1	0.23	A
1	0.31	D11
1	0.32	D4
1	0.78	Log P
1	0.81	(Log $P)^2$
2	0.37	D11, A
2	0.53	D4, A
2	0.57	D4, D11
2	0.79	Log P, D11
2	0.81	(Log $P)^2$, D11
2	0.84	Log P, (log $P)^2$
2	0.86	Log P, A
2	0.87	(Log $P)^2$, A
2	0.89	Log P, D4
2	0.90	(Log $P)^2$, D4
3	0.64	D4, D11, A
3	0.84	Log P, (log $P)^2$, D11
3	0.87	Log P, (Log $P)^2$, A
3	0.88	Log P, D11, A
3	0.88	(Log $P)^2$, D11, A
3	0.90	Log P, D4, D11
3	0.92	(Log $P)^2$, D4, D11
3	0.92	Log P, (log $P)^2$, D4
3	0.92	Log P, D4, A
3	0.93	(Log $P)^2$, D4, A
4	0.88	Log P, (log $P)^2$, D11, A
4	0.93	Log P, (log $P)^2$, D4, D11
4	0.93	Log P, (log $P)^2$, D4, A
4	0.94	Log P, D4, D11, A
4	0.95	(Log $P)^2$, D4, D11, A
5	0.95	Log P, (log $P)^2$, D4, D11, A

D. INTERPRETATION OF THE QSAR RESULTS

The R^2 values listed in Table 9.7 suggest that, in terms of explained variation, $(\log P)^2$ is slightly superior to log P. In terms of a physical model, however, an equation in $(\log P)^2$ but not log P would mean that there is an optimum or minimum potency at log $P = 0.0$. Figure 9.5 shows that this is well outside the range of log P values in the data set, for which the lowest value is 2.58. Because we also know that log P and $(\log P)^2$ are essentially totally correlated, the best interpretation of the data is in terms of log P rather than $(\log P)^2$. This discussion illustrates why it is so important to include plots in the initial analysis of a data set.

From the R^2 of Model 9.1 (Table 9.8), we see that 79% of the variation in potency is explained by variation in log P. Thus, if we wished to predict the potency of analogs that were varied at positions other than those examined in this data set, Equation 9.1 would be used. If only erythromycin A or B analogs were in the new data set, then Equation 9.2, which explains 86% of the variation in potency in the original data set, would be used. However, the equations may not make accurate predictions of analogs whose log P is outside the range used to generate the equation or that are substituted at other positions. Such predictions might be considered to be provisional tests of the generality of the equation and to provide some information that may help the synthetic chemist decide which of several analogs to make.

In terms of R^2, Model 9.6 is the optimum. However, from Table 9.6, neither the log P nor the $(\log P)^2$ term is significant. Of the equations that include only significant terms, that with the highest R^2 is Model 9.5. However, note that Model 9.4 explains only 2% less of the variation in log $(1/C_{SA})$. Thus, one should wonder if Model 9.5 is really "better than" Model 9.4.

TABLE 9.8
Selected QSAR Equations to Fit Log(1/C_SA) of Erythromycin Esters

Model	Coefficient (standard deviation)						Statistics	
	Intercept	Log P	(Log P)²	D4	D11	A	R²	s
9.1	6.21	−1.25	—	—	—	—	0.786	0.237
	(0.41)	(0.13)						
9.2	7.75	−1.70	—	—	—	−0.46	0.864	0.193
	(0.52)	(0.16)				(0.12)		
9.3	6.01	−1.12	—	−0.35	—	—	0.889	0.174
	(0.30)	(0.10)		(0.07)				
9.4	7.15	−1.47	—	−0.28	—	−0.33	0.925	0.146
	(0.42)	(0.13)		(0.06)		(0.10)		
9.5	6.89	−1.36	—	−0.29	−0.17	−0.36	0.945	0.127
	(0.89)	(0.12)		(0.05)	(0.06)	(0.08)		
9.6	4.56	0.07	−0.22	−0.29	−0.16	−0.30	0.948	0.128
	(2.53)	(1.55)	(0.24)	(0.06)	(0.06)	(0.11)		

The discussion of the "best" regression equation for a given set of data emphasizes the subjective nature of this choice. Which model is best depends on the use one wishes to make of it (predict new analogs versus test a model versus summarize data), the investigator's evaluation of the importance of parsimony or generalization versus precision and detail, as well as conscious or unconscious biases of the particular investigator.

E. FOLLOW-UP OF THE EQUATION

The regression analysis suggests that hydrophobicity is the prime determinant of potency and that the optimum partition coefficient is equal to or lower than the least hydrophobic analog tested. Thus, less hydrophobic esters might be more potent.

Parenthetically, it may be noted that the potency of these same analogs versus the Gram-negative bacterium *Klebsiella pneumoniae* was also a negative function of log P. It was our hope that the optimum log P would be lower for anti-Gram-negative activity. This would mean that we might find an erythromycin analog with increased anti-Gram-negative properties. The results of our further studies will be presented in Chapter 10.

IV. LESSONS LEARNED

Prior to generating the QSAR equations, inspection of the univariate correlations between potency and individual molecular properties and between the individual properties can provide guidance for the later studies. For example, if two molecular properties are highly correlated, then it is likely that no equation will include both as statistically significant. The strategy for discovering the "best" regression equation for a 2D QSAR requires the consideration of many possible equations. Although detailed information on all possible equations is not needed, summary information such as the R^2 or s from each possible equation can provide guidance. Both forward and backward selection procedures are helpful, but they must be followed with more detailed inspection of related equations.

REFERENCES

1. Martin, Y. C.; Jones, P. H.; Perun, T.; Grundy, W.; Bell, S.; Bower, R.; Shipkowitz, N. *J. Med. Chem.* **1972**, *15*, 635–8.
2. Hansch, C.; Steward, A. R.; Iwasa, J. *Mol. Pharmacol.* **1965**, *14*, 87–92.
3. Hansch, C.; Leo, A.; Hoekman, D. *Exploring QSAR: Hydrophobic, Electronic, and Steric Constants.* American Chemical Society: Washington, DC, 1995.
4. Perun, T. J. In *Drug Action and Drug Resistance in Bacteria*; Mitsuhashi, S., Ed. Univ. Park Press: Baltimore, MD, 1971, pp. 123–52.
5. Iwasaki, H.; Sugawara, Y.; Adachi, T.; Morimoto, S.; Watanabe, Y. *Acta Crystallogr., Sect. C: Cryst. Struct. Commun.* **1993**, *49*, 1227–30.
6. CLOGP, version 4.3. Biobyte Corporation. Claremont, CA. 2007. http://biobyte.com/bb/prod/40manual.pdf (accessed Nov. 12, 2007).
7. LogKow/KowWin, Syracuse Research Corporation. Syracuse. 2007. http://www.syrres.com/esc/est_kowdemo.htm (accessed Nov. 1, 2007).
8. Microsoft® Excel 2000, Microsoft. Redmond, WA. 2000.
9. SAS, version 8.0. SAS Institute Inc. Cary, NC.

10 Case Studies

The case studies discussed in this chapter illustrate the types of problems on which QSAR and molecular modeling was used as part of a drug discovery team at Abbott Laboratories. The first four case studies are of physical property-based and Free–Wilson 2D QSARs from the 1970s and 1980s; the next one is of molecular modeling and 3D QSAR from the 1980s; and the final one illustrates the use of 3D QSAR with structures from ligand–protein complexes. For the 2D QSARs, two analyses will be presented for each set of analogs: The first is of the data and conclusions as reported originally. The second highlights the additional insights that can be gained by reexamining the data with newer methods.

I. INHIBITION OF DOPAMINE β-HYDROXYLASE BY 5-SUBSTITUTED PICOLINIC ACID ANALOGS

A. EARLY QSARS
(Collaborators: E. B. Chappell, K. R. Lynn, P. H. Jones, J. Kyncl, F. N. Minard, and P. Somani)

Dopamine β-hydroxylase, EC 1.14.17.1, is a copper-containing enzyme that catalyzes the conversion of dopamine (Structure 10.1) to norepinephrine (Structure 10.2) (Figure 10.1). Because norepinephrine raises blood pressure but dopamine does not, we were interested in inhibitors of this enzyme for use as potential antihypertensives.

The literature reported that blood pressure is lowered when the animals are treated with alkyl analogs of fusaric acid (Structure 10.3, R = n-C_4H_9),[1] an inhibitor of dopamine β-hydroxylase. Our goal was to use the result of physical property-based 2D QSAR of analogs represented by the variation in the R-group of Structure 10.3 to optimize a novel inhibitor of the enzyme. In particular, we assumed that the physical properties responsible for the relative potency of these analogs would also be important in other series of inhibitors.

We first studied the straight-chain analogs listed in Table 10.1. We found that the negative log of the concentration that inhibits the enzyme 50%, log(1/C), is a parabolic function of the π value of the R substituent, Model 10.1 in Table 10.2. The optimum π value is 2.47. Model 10.1 reasonably predicts the inhibitory potency of the four subsequently reported[2] metabolites of fusaric acid, which are also included in Table 10.1.

The next literature report described the inhibitory potencies of 14 analogs in which the side chain was branched or incorporated a halogen or a phenyl ring (Table 10.3).[3] Because log (1/C) for fusaric acid was different in this publication from that reported earlier, we calculated a Model 10.2 for these analogs only. The new model is also parabolic in π and the optimum π, 2.40, is very similar to that of Model 10.1.

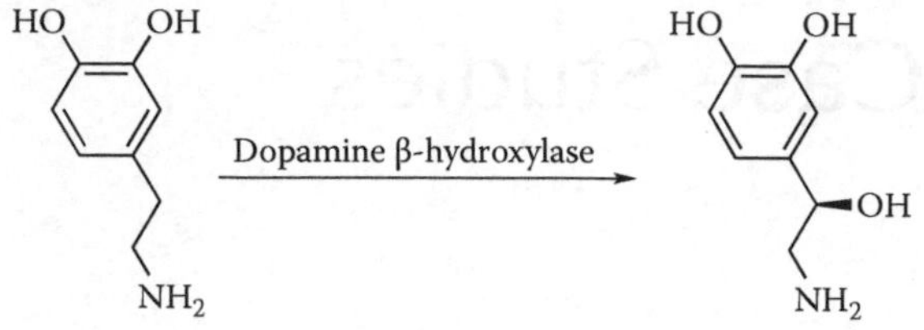

Dopamine
10.1

Norepinephrine
10.2

FIGURE 10.1 The reaction catalyzed by dopamine β-hydroxylase.

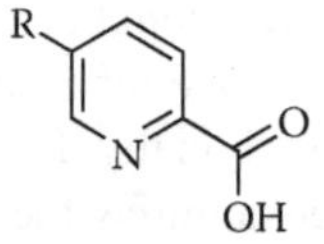

(10.3)

TABLE 10.1
Structure, π Values, and Potency of Analogs of Fusaric Acid (Structure 10.3, R = C_4H_9)

Structure of R	π	Log (1/C) Obs.	Log (1/C) Calc. Model 10.1
	Molecules Used to Derive Model 10.1[a]		
CH_3	0.5	5.40	5.71
C_2H_5	1.0	5.70	6.24
n-C_3H_7	1.5	6.52	6.62
n-C_4H_9	2.0	7.05	6.84
n-C_5H_{11}	2.5	7.30	6.92
n-C_6H_{13}	3.0	6.90	6.83
n-C_7H_{15}	3.5	6.60	6.60
n-C_8H_{17}	4.0	6.10	6.20
n-C_9H_{19}	4.5	5.60	5.66
	Metabolites Used to Test Model 10.1[b]		
$(CH_2)_3C(=O)OH$	0.85	6.00	6.10
$(CH_2)_2CH(OH)CH_3$	0.64	5.70	5.87
$CH_2CH(OH)C_2H_5$	0.64	6.15	5.87
$(CH_2)_4OH$	0.84	5.82	6.09

[a] Reported by Suda, H.; Takeuchi, T.; Nagatsu, T.; Matsuzaki, M.; Matsumoto, I.; Umezawa, H. *Chem. Pharm. Bull.* **1969**, *17*, 2377–80.

[b] Structures and potency reported by Umezawa, H., *Enzyme Inhibitors of Microbial Origin.* University of Tokyo Press: Tokyo, 1972.

TABLE 10.2
Property-Based 2D QSAR Models of the Potency of Fusaric Acid Analogs

$$\text{Log}(1/C) = a + b \times \pi - c \times \pi^2$$

Structure of R	Source of Log (1/C)	a	b	c	Optimum π	R^2	n	s	Model Number
Unsubstituted and n-alkyl	Table 10.1	5.02	1.53	−0.31	2.47	0.81	10	0.34	10.1
Halogenated and branched alkyl	Table 10.3	4.58	3.03	−0.63	2.40	0.88	14	0.19	10.2
Unsubstituted; alkyl; halo, hydroxyl, and phenyl alkyl	Table 10.3	−1.48	1.25	−0.27	2.31	0.67	28	0.56	10.3

Lastly, the potencies of an even larger set of analogs was reported.[4] Because log $(1/C)$ for fusaric acid varied by tenfold in several publications, we could not use the older models to predict the potency of the new analogs. In addition, the relative potency of some of the analogs also changed. Even though we did not understand the reasons for the changes in potency, we continued to model the structure–activity relationships. Thus, we calculated Model 10.3 using as the dependent variable log $(1/C)$ corrected for the potency of fusaric acid in the particular report. The structures, physical properties, and corrected log $(1/C)$ values are collected in Table 10.3.

Model 10.3 was fitted using the potencies of the unsubstituted, alkyl, phenyl alkyl, halo alkyl, and hydroxyl alkyl analogs. Although it is still statistically significant, Model 10.3 has a much lower R^2 and higher s than do Models 10.1 and 10.2. Thus, properties in addition to hydrophobicity appear to affect the potency of the more diverse analogs. Nonetheless, the original model based on 11 alkyl analogs is very similar to the one based on 28 more diverse analogs, and it was sufficient to predict the relative potency of the more complex metabolites and newer analogs. In addition, the optimum π value is within 0.1 of 2.4 for the three models.

B. Reexamination of the QSAR

In this work we standardized on the potency values listed in Table 10.3. We also described the hydrophobicity of the molecules using log$P$ values calculated with KowWin.[5] These calculated log$P$ values are perfectly correlated with the original π values of the n-alkyl substituted analogs, but differ somewhat for the other molecules.

Model 10.4 (Table 10.4) shows that using the new potency values for the 10 analogs that led to Model 10.1 decreases the significance of the fit to the 2D QSAR equation. However, the fit is improved if the two branched alkyl molecules are included

TABLE 10.3

Structure, π, Log P[5], and Relative Potency of the Whole Set of Fusaric Acid Analogs

Structure of R	Included in Models	π	Log P	Log (1/C) Obs.	Calculated Model 10.3	10.8	10.11
$(CH_2)_2CH(Br)$ CH_2Br	10.2, 10.3, 10.8, 10.9, 10.11, 10.12	2.49	3.32	0.88	−0.03	0.73	1.13
$(CH_2)_2CH(Br)CH_3$	10.2, 10.3, 10.8, 10.9, 10.11, 10.12	1.89	2.98	0.88	−0.08	0.48	0.18
$(CH_2)_2CH(Cl)$ CH_2Cl	10.2, 10.3, 10.8, 10.9, 10.11, 10.12	2.31	3.14	0.88	−0.03	0.64	1.08
$(CH_2)_4Cl$	10.2, 10.3, 10.8, 10.9, 10.11, 10.12	1.92	2.96	0.38	−0.07	0.24	0.35
$(CH_2)_5Cl$	10.2, 10.3, 10.8, 10.9, 10.11, 10.12	2.42	3.45	0.34	−0.03	0.09	0.02
$(CH_2)_4Br$	10.2, 10.3, 10.8, 10.9, 10.11, 10.12	2.09	3.05	0.24	−0.04	0.24	0.27
$(CH_2)_2CH(CH_3)_2$	10.2, 10.3,10.5–10.12	2.30	3.12	0.21	−0.03	0.58	0.12
$(CH_2)_2CH(Cl)CH_3$	10.2, 10.3, 10.8, 10.9, 10.11, 10.12	1.72	2.89	0.16	−0.13	0.39	0.03
$CH_2CH(CH_3)_2$	10.2, 10.3,10.5–10.12	1.80	2.63	−0.11	−0.10	−0.44	−0.50
$(CH_2)_6Cl$	10.2, 10.3, 10.8, 10.9, 10.11, 10.12	2.92	3.94	−0.13	−0.12	−0.39	−0.04
$(CH_2)_5F$	10.2, 10.3, 10.8, 10.9, 10.11, 10.12	1.77	3.14	−0.17	−0.11	−0.16	−0.29
$(CH_2)_4F$	10.2, 10.3, 10.8, 10.9, 10.11, 10.12	1.27	2.65	−0.38	−0.33	0.09	−0.08
$(CH_2)_3Br$	10.2, 10.3, 10.8, 10.9, 10.11, 10.12	1.59	2.56	−0.49	−0.17	−0.23	−0.47
$(CH_2)_3Cl$	10.2, 10.3, 10.8, 10.9, 10.11, 10.12	1.42	2.47	−0.60	−0.25	−0.34	−0.39
n-C_5H_{11}	10.3–10.12	2.50	3.20	0.19	−0.03	0.18	−0.32
n-C_4H_9	10.3–10.12	1.98	2.71	0.00	−0.06	−0.01	−0.71
n-C_6H_{13}	10.3–10.12	3.00	3.69	−0.17	−0.15	0.13	0.21
n-C_7H_{15}	10.3–10.12	3.50	4.18	−0.40	−0.39	−0.50	−0.37
n-C_3H_7	10.3–10.12	1.53	2.22	−0.57	−0.20	−0.71	−0.75
n-C_8H_{17}	10.3–10.12	4.00	4.67	−0.96	−0.78	−1.03	−1.00
C_2H_5	10.3–10.12	1.02	1.72	−1.40	−0.49	−1.34	−1.38
n-C_9H_{19}	10.3–10.12	4.50	5.16	−1.40	−1.29	−1.27	−1.21
CH_3	10.3–10.12	0.56	1.23	−1.80	−0.87	−1.43	−1.35
H	10.3, 10.4, 10.8, 10.9, 10.11,10.12	0.00	0.69	−1.23	−1.48	−1.60	−1.28
$CH_2CH(OH)C_2H_5$	10.3, 10.8, 10.9, 10.11, 10.12	1.17		−0.90		0.07	−0.80

(Continued)

TABLE 10.3 (CONTINUED)
Structure, π, Log P[5], and Relative Potency of the Whole Set of Fusaric Acid Analogs

| | | | | | Log (1/C) | | |
| | | | | | Calculated Model | | |
Structure of R	Included in Models	π	Log P	Obs.	10.3	10.8	10.11
(CH$_2$)$_4$OH	10.3, 10.8, 10.9, 10.11, 10.12		1.24	−1.23		0.03	−0.88
(CH$_2$)$_2$CH(OH)CH$_3$	10.3, 10.8, 10.9, 10.11, 10.12		1.17	−1.35		0.27	−0.47
(CH$_2$)$_5$C$_6$H$_5$	10.3, 10.8, 10.9, 10.11, 10.12	4.1	4.91	−1.60	−0.87	−0.50	−1.64
(CH$_2$)$_3$C(=O)OH	10.8, 10.9, 10.12		1.47	−1.05		−0.00	−1.05
NHC(=S)N(CH$_3$)$_2$	10.8, 10.9, 10.12	1.59	0.32	−1.12	−0.17	0.06	−1.38
CH$_2$OC(=O)N(CH$_3$)$_2$	10.8, 10.9, 10.12	0.23	0.77	−1.89	−1.21	0.11	−1.93
SO$_2$NHCH$_3$	10.8, 10.9, 10.12	−1.4	−0.47	−3.00	−3.74	−0.60	−2.91

TABLE 10.4
Models that Relate Log P to Potency[4] of Alkyl Fusaric Acid Analogs ($n = 11$ and Whole Data set, $n = 32$)

Structure of R	Equation	a	b	c	n	R^2	s	Model Number
Unsubstituted and n-alkyl	I	−2.96	1.73	−0.27	10	0.66	0.45	10.4
Branched and n-alkyl	I	−5.02	3.067	−0.464	11	0.95	0.18	10.5
Branched and n-alkyl	II	—	73.18	73.30	11	0.57	0.50	10.6
Branched and n-alkyl	III	0.90	1.59	2.67	11	0.98	0.11	10.7
All	III	0.748	1.06	2.04	32	0.57	0.60	10.8
All	I	−2.57	1.67	−0.27	32	0.69	0.45	10.9

I. $\log(1/C) = a + b \times \log P - c \times (\log P)^2$
II. $\log(1/C) = a + b \times (\log P) - c \times \log(10^{(\log P)} + 1) = a + b \times \log P - c \times \log(P + 1)$
III. $\log(1/C) = a + b \times (\log P - 3) - c \times \log(10^{(\log P - 3)} + 1)$

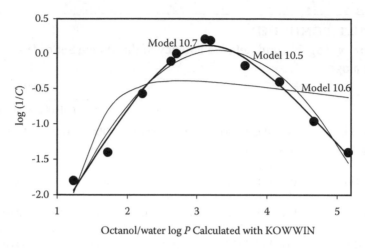

FIGURE 10.2   A plot of log P calculated with KowWin versus log $(1/C)$ for the alkyl analogs of fusaric acid and the curves for the models that fit this data.

and the unsubstituted analog is omitted because it might have different electronic properties (Model 10.5).

We next explored fits to model-based equations that relate potency to log P. Comparing Models 10.6 and 10.7 shows that although a fit to a model-based equation was not possible using the log P values directly, a successful fit was obtained when the value of the approximate optimum, as seen in Figure 10.2, was subtracted from log P. This correction appears to be necessary because, in fitting Model 10.6, $P \gg 1.0$ and hence dominates the sum in the third term. The fit to Model 10.7 is somewhat more precise than that to the parabolic 2D QSAR Model, 10.5. Figure 10.2 shows that this fit is closer for the four most potent analogs. However, Model 10.8, which includes all 32 analogs, fits the data less well than does Model 10.9, a parabolic 2D QSAR model. The similarity of Model 10.9 to Model 10.4 suggests again that factors other than calculated log P influence potency.

We next considered the possibility that the structure–activity relationships may be governed by steric rather than hydrophobic effects, that is, that CoMFA steric fields (2 Å spacing, Sybyl defaults) might also fit the data. Table 10.5 shows that the fit to the 3D QSAR CoMFA, Model 10.10, is of approximately the same precision as the fit to the corresponding relationship in log P, Model 10.7. Comparing Models 10.3 and 10.11, which are based on 28 analogs, we see that the fit to the CoMFA 3D QSAR is superior. The same is true for models that contain all the analogs: Models 10.9 versus 10.12. Figure 10.3 shows how the combination of positive and negative steric fields explains the observed dependence of potency on the size of the substituent.

Table 10.3 lists the observed potency and that calculated by several of the models, and Figure 10.4 shows the relationships between observed and fitted values for several of the equations. From these plots and Models 10.11 and 10.12, one can see that CoMFA 3D QSAR steric fields are more successful than log P at extending the model to contain all of the analogs.

TABLE 10.5
Summary of CoMFA Analyses of Fusaric Acid Derivatives Using Steric Fields

Structure of R	Number of Latent Variables	n	Fitted		Cross-Validated		Model Number
			R^2	s	q^2	s	
Straight and branched alkyl	2	11	0.97	0.13	0.82	0.34	10.10
Unsubstituted, alkyl, halo, hydroxyl, and phenyl alkyl	2	28	0.88	0.26	0.79	0.35	10.11
All	3	32	0.87	0.34	0.39	0.74	10.12

Adding electrostatic fields to the CoMFA analysis did not improve the fits or cross-validation, presumably because there are so few polar analogs in the data set. However, others analyzed a different data set, one with more varied substituents on the pyridine ring, and found that in addition to steric effects it was also necessary to include electrostatic effects in both 2D QSAR and CoMFA 3D QSAR.[6]

C. LESSONS LEARNED FROM THE ANALYSIS OF FUSARIC ACID ANALOGS

1. Model-based equations based on log P fit a subset of the data better than does a parabolic one. However, the parabolic equation fits the whole data set somewhat more precisely. Both support the hypothesis that the hydrophobicity of the molecule is a primary determinant of its potency.
2. 2D QSAR and the 3D QSAR method CoMFA 3D QSAR lead to different conclusions as to the physical properties associated with inhibitory potency.

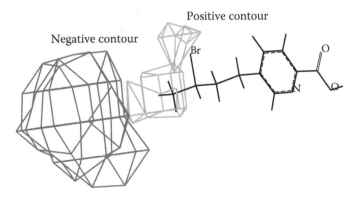

FIGURE 10.3 The positive and negative CoMFA 3D QSAR steric fields calculated by Model 10.12.

This ambiguity arises from the accidental high correlation between hydro-phobicity and steric properties. However, the CoMFA 3D QSAR steric prop-erties fit the varied analogs more precisely than does a parabolic function in log P.

3. The overall picture that emerges from these studies is that the binding site on dopamine β-hydroxylase for the 5-substituent of fusaric analogs is non-polar and limited in size.

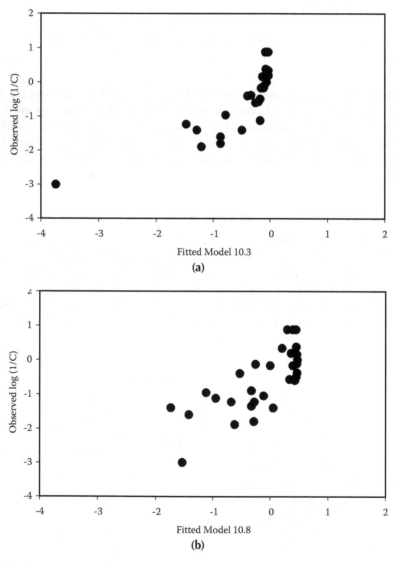

FIGURE 10.4   Plots of observed versus fitted potency of fusaric acid analogs. (a) Fits to Model 10.3, (b) fits to Model 10.8, and (c) fits to Model 10.11.

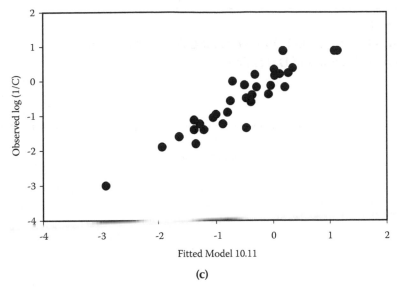

FIGURE 10.4 Continued

II. THE RATE OF HYDROLYSIS OF AMINO ACID AMIDES OF DOPAMINE

A. EARLY QSARS
(Collaborators: P. H. Jones, J. Kyncl, K. R. Lynn, F. N. Minard, C. W. Ours, and P. Somani)

Dopamine increases the blood flow to the kidney, a desirable therapeutic effect.[7] However, it must be administered intravenously, and this leads to the undesirable effects of increased blood pressure and heart rate. Therefore, we sought to discover analogs of dopamine that could be administered orally but would release dopamine selectively in the kidney. This work focused on analogs in which the nitrogen of dopamine has been converted into an amide of an amino acid, Structure 10.4,[8] because there is an α-aminoacyl amidase in kidney tissue that hydrolyzes such compounds.

The relative rates of hydrolysis of the amides and their calculated physical properties are reported in Table 10.6. The log P value is that measured for the amino acid. Because the free amino group might be involved in binding to the enzyme, we used estimated σ^* values to represent the substituent effects on its pK_a.

(10.4)

TABLE 10.6
Physical Properties and Observed and Calculated Hydrolysis Rates of Amino Acid Amides of Dopamine

| | | | | | | | Log (1/k) | | |
| | | | | | | | | Calc. Model | |
Amide	R in Structure 10.4	Log P	σ^*	D1	D2	Obs.	10.13	10.14	10.15
	Analogs Used to Derive Models 10.13–10.15								
Alanine	-CH(NH$_2$)CH$_3$	−2.94	0.88	0.0	0.0	3.00	2.08	2.40	2.63
Phenylalanine	-CH(NH$_2$)CH$_2$C$_6$H$_5$	−1.43	1.06	0.0	0.0	2.81	2.36	2.40	2.40
α-amino butyric acid	-CH(NH$_2$)CH$_2$CH$_3$	−2.38	0.86	0.0	0.0	2.79	2.05	2.40	2.65
Glycine	-CH$_2$NH$_2$	−3.03	0.98	0.0	0.0	2.26	2.23	2.40	2.52
Leucine	-CH(NH$_2$)CH$_2$CH(CH$_3$)$_2$	−1.71	0.85	0.0	0.0	2.18	2.03	2.40	2.65
Serine	-CH(NH$_2$)CH$_2$OH	−4.10	1.18	0.0	0.0	1.95	2.55	2.40	2.15
Aspartic acid	-CH(NH$_2$)CH$_2$C(=O)OH	−3.59	1.30	0.0	0.0	1.84	2.74	2.40	1.83
Valine	-CH(NH$_2$)CH(CH$_3$)$_2$	−2.10	0.79	1.0	1.0	0.84	0.86	0.66	0.85
Isoleucine	-CH(NH$_2$)CH(CH$_3$)C$_2$H$_5$	−1.61	0.77	1.0	1.0	0.84	0.83	0.66	0.85
Proline	-*cyclo*-CH$_2$NHCH$_2$CH$_2$CH$_2$-	−2.10	0.23	0.0	1.0	0.30	1.06	0.66	0.29
	True Predictions: Analogs Synthesized after Models 10.13–10.15 Were Developed								
Threonine	-CH(NH$_2$)CH(OH)CH$_3$		1.16	1.0	1.0	1.78	0.85[a]	0.66[a]	0.31
Tyrosine	-CH(NH$_2$)CH$_2$-C$_6$H$_4$-*p*-OH		1.10	0.0	0.0	2.54	2.40[a]	2.40[a]	2.32[a]
Pyroglutamic acid	-*cyclo*-CHNHC(=O)CH$_2$CH$_2$		1.70	0.0	1.0	<0.30	−0.76[a]	0.66[a]	−1.59[a]
Sarcosine	-CH$_2$NHCH$_3$		0.49	0.0	0.0	<0.30	1.75	2.40	0.74
N-methylalanine	-CH(NHCH$_3$)CH$_3$		0.39	0.0	1.0	<0.30	1.27	0.66[a]	0.61[a]
α-Methylalanine	-C(CH$_3$)(NH$_2$)CH$_3$		0.79	1.0	1.0	<0.30	0.85[a]	0.66[a]	0.85[a]
β-Alanine	-(CH$_2$)$_2$NH$_2$		0.48	0.0	0.0	0.70	1.71	2.40	2.57
γ-Aminobutyric acid	-(CH$_2$)$_3$NH$_2$		0.23	0.0	0.0	<0.30	1.06[a]	2.40	2.12

[a] The observed value is within the 95% confidence interval of the predicted value.

There is no statistically significant model that fits log k with log P and σ^*. Three analogs deviate from apparent relationship between log k and σ^*: the prolyl, valyl, and isoleucyl amides. To account for structural differences between these analogs and the others, we investigated two indicator variables. The first, D1, represents

TABLE 10.7

Models that Relate Molecular Properties to the Rate of Enzymatic Hydrolysis of Dopamine Amides

$\log k = -1.57(\pm0.29) - 1.70(\pm0.20)D1 + 9.34(\pm1.60)\sigma^* - 5.21(\pm1.04)\sigma^{*2}$ 10.13

$R^2 = 0.917, s = 0.328, n = 10, \sigma^*_{opt} = 0.90$

$\log k = 2.40 - 1.74D2$ 10.14

$R^2 = 0.814, s = 0.426, n = 10$

$\log k = 1.45(\pm0.78) - 1.84(\pm0.32)D2 + 3.55^a(\pm1.56)\sigma^* - 2.50(\pm0.96)(\sigma^*)^2$ 10.15

$R^2 = 0.915, \sigma = 0.332, n = 10, \sigma^*_{opt} = 0.71$

a Significant at 6% level.

branching at the β carbon: It is set equal to 1.0 for the valyl and isoleucyl analogs and 0.0 for the remaining analogs. Model 10.13, Table 10.7, indicates an optimum σ* of 0.90, close to the value of the alanyl and glycyl analogs, and a large negative dependence of hydrolysis rate on substitution at the β carbon. The second indicator variable was set equal to 1.0 for the prolyl, valyl, and isoleucyl amides. It leads to Models 10.14 and 10.15. Again, a large negative dependence on the indicator variable and an optimum σ* value is suggested.

Table 10.6 lists the predicted and observed hydrolysis rates of eight additional amides that were synthesized after the models were developed. Models 10.13 and 10.14 correctly predict the rate for five of the eight new analogs, and Model 10.15 correctly predicts four. In particular, the inactive analogs were correctly predicted only for those models that use the more or less arbitrary indicator variable. We concluded that the models are not predictive enough to be useful.

During the course of this work, the project biochemist discovered that the intestine is rich in the same aminoacyl amidase.[9] Hence, none of the analogs could meet the goal of kidney-specific release of dopamine because they would be hydrolyzed in the gut before they reached the general circulation. However, a quick literature survey revealed that γ-glutamyl amidase is located primarily in the kidney. We therefore suggested the synthesis of the γ-glutamyl amide of dopamine, for which $-(CH_2)_2CH(NH_2)C(=O)OH$. It was found to be selectively hydrolyzed in the kidney to produce increases in renal blood flow with minimal effects on blood pressure and heart rate.[10–12] Although the compound met the objectives of the discovery team, commercial concerns prevented its further development.

B. REEXAMINATION OF THE QSAR

We investigated the possibility that newer QSAR methods might uncover a significant relationship between molecular properties and hydrolysis rate. However, there is no significant linear or nonlinear relationship between the log k for hydrolysis by the enzyme and log P calculated with CLOGP[13] or KowWin,[5] pK_a's calculated with PharmaAlgorithm software,[14] and the indicator variables. Additionally, no significant CoMFA 3D QSAR[15] steric or steric plus electrostatic models were found with 3D structures generated with CONCORD[16] and superimposed either on the dopamine substructure or on the amide and basic amino group.

C. Lessons Learned from the Analysis of Amino Acid Pro-Drugs of Dopamine

1. Some data sets do not produce robust QSARs.
2. The indicator variables were arbitrarily chosen.
3. Careful examination of the biological properties of the analogs led to the suggestion of an alternative pro-drug that met the goals of the project.

III. ANALGETIC POTENCY OF γ-CARBOLINES

A. Early QSARs
(Collaborators: P. W. Dodge, A. T. Dren, D. Garmaise, R. P. Johnson, E. T. Kimura, and P. R. Young)

The aim of this work was to find a potent tranquilizer with the generic Structure 10.5, a γ-carboline. Because the analogs showed both tranquilizing and analgetic activity but analgetic activity is more straightforward to measure, optimization was based on the ED_{50} in a test for analgesia. Table 10.8 lists the structures and potency of the original analogs.[17]

To investigate possible QSAR models, we considered the properties in Table 10.8. In addition, we investigated log P calculated using the sum of π values for the X and Y substituents; individual π values of the substituents at position 8 of the carboline ring and the para position of a phenyl in Y; an indicator variable for the presence or absence of a ketone in Y; and $\Sigma\sigma_R$ and $\Sigma\sigma_I$. The $\Sigma\sigma_R$ and $\Sigma\sigma_I$ values represent the resonance and inductive effects of carboline ring substituents on the charge on the indole nitrogen.[18] $E_{s,8}$ is the Taft steric value[19] for the substituent at position 8 and $E_{s,p}$ that for the para position of any aromatic ring at position Y.

Preliminary models predicted much higher than observed potency of the 8-Cl-*p*-F-butyrophenone analog. We omitted it from further analysis because the observed ED_{50} has a wide confidence interval that overlaps the predicted, and its low potency did not warrant retesting.

Table 10.9 lists the statistically significant QSAR models. The models suggest that the relative potency of these analogs is governed by differences in steric and electronic properties. In particular, Models 10.17 and 10.18 indicate a positive steric

(10.5)

TABLE 10.8

Structure, Physical Properties, and Analgetic Potency of γ-Carbolines (Structure 10.5)

X	Y	$\Sigma\sigma_R$	$E_{s,8}$	$E_{s,p}$	Obs.	Calc. Model 10.16
					Log $(1/ED_{50})$	
	Analogs Used to Generate Models 10.16–10.19					
5 CH$_3$, 8F	(CH$_2$)$_3$C(=O)C$_6$H$_4$-p-F	−0.32	−0.46	−0.46	1.74	1.50
6 CH$_3$	(CH$_2$)$_3$C(=O)C$_6$H$_4$-p-F	−0.14	0.00	−0.46	1.92	1.71
8 Br	(CH$_2$)$_3$C(=O)C$_6$H$_4$-p-F	−0.16	−1.10	−0.46	1.38	1.42
8 CF$_3$	(CH$_2$)$_3$C(=O)C$_6$H$_4$-p-F	0.61	−2.40	−0.46	1.49	1.48
8 CH$_3$	(CH$_2$)$_3$C(=O)C$_6$H$_4$-p-F	−0.14	−1.20	−0.46	1.37	1.40
8 Cl	H	−0.18	−0.98	0.00	1.13	1.04
8 CN	(CH$_2$)$_3$C(=O)C$_6$H$_4$-p-F	1.00	−0.55	−0.46	2.17	2.16
8 F	(CH$_2$)$_3$C(=O)C$_6$H$_5$	−0.32	−0.46	0.00	1.44	1.10
8 F	(CH$_2$)$_3$CH(OH)C$_6$H$_4$-p-F	−0.32	−0.46	−0.46	1.45	1.50
8 F	(CH$_2$)$_3$CN	−0.32	−0.46	0.00	0.80	1.10
8 F	(CH$_2$)$_3$OC$_6$H$_4$-p-F	−0.32	−0.46	−0.46	1.59	1.50
8 F	(CH$_2$)$_4$C(=O)C$_6$H$_4$-p-F	−0.32	−0.46	−0.46	1.42	1.50
8 F	(CH$_2$)$_4$C$_6$H$_4$-p-F	−0.32	−0.46	−0.46	1.53	1.50
8 F		−0.32	−0.46	−0.55	1.43	1.58
8 F	CH$_2$C$_6$H$_5$	−0.32	−0.46	0.00	0.93	1.10
8 F	H	−0.32	−0.46	0.00	1.12	1.10
8 OCH$_3$	(CH$_2$)$_3$C(=O)C$_6$H$_4$-p-F	−0.43	−0.55	−0.46	1.40	1.42
H	(CH$_2$)$_3$C(=O)C$_6$H$_4$-p-F	0.00	0.00	−0.46	1.63	1.79
	Analogs Used to Test Models 10.16–10.19					
8 Cl	(CH$_2$)$_3$C(=O)C$_6$H$_4$-p-F	−0.18	−0.98	−0.46	0.90[a]	1.43
8 F	(CH$_2$)$_3$C(=O)C$_6$H$_4$$p$-F	−0.32	−0.46	−0.46	2.18[a]	1.50
8 F	(CH$_2$)$_4$C$_6$H$_4$-p-NO$_2$	−0.32	−0.46	−2.52	<0.50[b]	3.27
8 F	(CH$_2$)$_3$C(=O)C$_6$H$_4$-p-C(CH$_3$)$_3$	−0.32	−0.46	−2.78	0.80[b]	3.49
8 F	(CH$_2$)$_3$C(=O)C$_6$H$_4$-p-NH$_2$	−0.32	−0.46	−0.61	1.09[b]	1.63
8 F	(CH$_2$)$_3$C(=O)C$_6$H$_4$-p-CH$_3$	−0.32	−0.46	−1.24	<0.50[b]	2.17
8 F	(CH$_2$)$_3$C(=O)C$_6$H$_4$-p-Cl	−0.32	−0.46	−0.97	<0.50[b]	1.94
8 F	(CH$_2$)$_3$C(=O)C$_6$H$_4$-p-NO$_2$	−0.32	−0.46	−2.52	<0.50[b]	3.27

[a] Not included in the calculation of Models 10.16 to 10.19.
[b] Synthesized to test Model 10.18.

TABLE 10.9

Structure–Activity Models for γ-Carbolines

Original Data Set

$$\log (1/ED_{50}) = 1.71 + 0.65 \sum \sigma_R + 0.28 E_{s,8} \qquad\qquad 10.16$$
$$R^2 = 0.46, s = 0.25, n = 18$$

$$\log (1/ED_{50}) = 1.19 + 0.35 \sum \sigma_R - 0.88 E_{s,p} \qquad\qquad 10.17$$
$$R^2 = 0.62, s = 0.21, n = 18$$

$$\log (1/ED_{50}) = 1.39 + 0.52 \sum \sigma_R - 0.86 E_{s,p} + 0.26 E_{s,8} \qquad\qquad 10.18$$
$$R^2 = 0.77, s = 0.17, n = 18$$

Analogs with *p*-*F*-butyrophenone Side Chain

$$\log (1/ED_{50}) = 1.83 + 0.48 \sum \sigma_R + 0.28 E_{s,8} \qquad\qquad 10.19$$
$$R^2 = 0.80, s = 0.15, n = 8$$

effect for the para substituent on Y. Note that the coefficient for $\sum \sigma_R$ in Models 10.18 and 10.19 are essentially identical for analogs in which Y is constant, Model 10.19, as for the larger data set, Model 10.18. The same is true for the coefficients of $E_{s,8}$.

Table 10.8 lists the observed and predicted potencies of eight analogs that were used to test the models. All are less potent than predicted.

B. REEXAMINATION OF THE QSAR

This data set is a challenge for QSAR because the standard deviation of the log $(1/ED_{50})$ values from the mean is only 0.34. This value is similar to deviations from the fit of successful QSAR models for other data sets, such as Models 10.1 and 10.12. Hence, it might be impossible to find a predictive model.

Model 10.18 is probably not a chance correlation even though its R^2 of 0.77 is based on exploring eight properties to explain the potency of 18 analogs. For example, the literature suggests that if one explores ten properties to explain the potency of 18 molecules, there is an approximately 0.06 chance of an R^2 greater than 0.7, and 0.02 chance of an R^2 greater than 0.8.[20] However, the chance for the example may be even lower because the structures (and hence properties) of the 18 analogs are not varied independently: nine of the original 18 analogs have fluoro as the X-substituent and ten have *p*-F-butryophenone as the Y-substituent.

At the time of the analysis, we hypothesized that the failure of the model to predict the potency of additional analogs could be traced to the $E_{s,p}$ descriptor. The *para*- substituents on the newly synthesized molecules are larger than any in the original data set and, thus, the predicted potencies are extrapolations from the original data. In addition, except for the cyclized analog, there are only two *para*- substituents in the original set: -H and -F. A further difficulty is that if the side chain does not contain an aromatic ring, we set $E_{s,p}$ to the same value as for analogs that contain an unsubstituted aromatic ring. Yet another problem is that $E_{s,p}$ does not account for the variation in the length of the chain between the carboline ring and the phenyl group.

In spite of the small variation in biological potency of these analogs, we investigated the possibility that newer QSAR methods might uncover a significant relationship. There is no significant linear or nonlinear relationship between pED_{50} and log P calculated with CLOGP[13] or KowWin,[5] log D's and pK_a's calculated with PharmaAlgorithm software,[14] and polar and nonpolar surface areas. Additionally, no significant CoMFA 3D QSAR[15] models were found with 3D structures generated with CONCORD in the extended conformation and superimposed on the carboline.

C. LESSONS LEARNED FROM THE ANALYSIS OF THE ANALGETIC PROPERTIES OF γ-CARBOLINES

1. Some data sets do not produce robust QSARs.
2. The $E_{0,8}$ predictor was arbitrarily chosen.
3. It can be a challenge to discover QSARs in data sets that have low variation in biological response and many repeats of the same substituent.

IV. ANTIBACTERIAL POTENCY OF ERYTHROMYCIN ANALOGS
(Collaborators: L. A. Frieberg, J. J. Hackbarth, W. Grundy, P. H. Jones, K. R. Lynn, J. C.-H. Mao, T. J. Perun, N. L. Shipkowitz, and A. M. von Esch)

Because it is effective against Gram-positive infections and not against Gram-negative infections, erythromycin A, Structure 1.9, cannot be used to treat Gram-negative infections. Hence, a goal of the analogs program was to discover a compound with greater efficacy against Gram-negative bacteria and also to increase potency against Gram-positive bacteria. Our hypothesis was that we could find clues as to how to broaden the antibacterial spectrum by exploring the relationship between the physical properties of the analogs and their antibacterial potency and spectrum. Two literature reports provided evidence of changes in bacterial spectrum with changes in log P: For both penicillin derivatives[21] and miscellaneous molecules,[22] the log P that corresponds to maximum potency is two to four log units lower for activity against Gram-negative than against Gram-positive bacteria.

The QSAR of alkyl erythromycin esters was examined in Chapter 9. Here we describe the preceding and additional studies that we made.

A. EARLY QSARs (1969)

1. Effect of Hydrophobicity on the Gram-Positive Antibacterial Activity of Erythromycin Esters

The compounds exert their antibacterial activity by inhibiting protein synthesis by binding to the 2058–2062 region of the 50S ribosomal RNA subunit.[23] Hence, we explored the possible relationship between physical properties and the ability of the esters in Table 10.10 to displace labeled erythromycin from isolated ribosomes.[24] However, we abandoned this effort because there is essentially no variation in this property for the nine esters that are not modified in the 2' position: The standard

TABLE 10.10
Potency of Erythromycin Esters against *S. aureus*

Substituents (Structures 1.9, 1.10)					Cell-Free % Erythromycin A Displaced	Log (1/C_{SA}) Obs.	Calc. Model 10.20
2'	4"	11	12	π			
-OH	-OH	-OH	-OH	0.0		3.00[a]	2.98
-OC(=O)H	-OH	-OH	-OH	0.39		2.91[a]	2.86
-OH	-OH	-OH	-H	1.80		2.82[a]	2.43
-OC(=O)H	-OH	-OH	-H	2.18		2.78	2.31
-OC(=O)CH₃	-OH	-OH	-OH	0.89	19	2.75[a]	2.71
-OH	-OC(=O)H	-OH	-OH	0.39		2.72	2.86
-OC(=O)H	-OC(=O)H	-OH	-OH	0.78		2.71	2.74
-OH	-OH	-OC(=O)H	-H	2.18	72	2.71[a]	2.32
-OC(=O)CH₃	-OC(=O)H	-OH	-OH	1.27	19	2.70	2.50
-OH	-OC(=O)H	-OH	-H	2.18		2.54	2.32
-OC(=O)CH₃	-OH	-OH	-H	0.89		2.54[a]	2.71
-OC(=O)H	-OC(=O)H	-OH	-H	2.58		2.50	2.20
-OH	-OC(=O)H	-OC(=O)H	-H	2.56	82	2.42	2.20
-OC(=O)H	-OC(=O)H	-OC(=O)H	-H	2.94		2.39	2.09
-OH	-OH	-OC(=O)CH₃	-H	2.69	65	2.30	2.16
-OC(=O)CH₃	-OC(=O)H	-OH	-H	3.07		2.28	2.05
-OH	-OC(=O)CH₃	-OH	-OH	0.89	79	2.24	2.70
-OC(=O)CH₃	-OC(=O)CH₃	-OH	-OH	1.78		2.15	2.44
vC(=O)C₆H₅	-OH	-OH	-OH	2.15		2.13[a]	2.32
-OC(=O)CH₃	-OH	-OC(=O)CH₃	-H	3.58	25	2.11	1.89
-OC(=O)CH₃	-OC(=O)CH₃	-OH	-H	3.58	9	2.04	1.89
-OH	-OC(=O)CH₃	-OH	-H	2.69	77	2.02	2.16
-OH	-OC(=O)C₂H₅	-OH	-H	3.18	82	2.02	2.01
-OC(=O)CH₃	-OC(=O)C₂H₅	-OH	-H	4.08	29	1.87	1.74
vC(=O)CH₃	-OC(=O)H	-OC(=O)CH₃	-H	3.96		1.65	1.77
-OH	-OC(=O)CH₃	-OC(=O)CH₃	-H	3.58	66	1.51	1.89
-OC(=O)CH₃	-OC(=O)CH₃	-OC(=O)CH₃	-H	4.47	30	1.39	1.62
-OH	-OC(=O)C₂H₅	-OC(=O)C₂H₅	-H	4.58	61	1.38	1.59
-OC(=O)CH₃	-OC(=O)C₂H₅	-OC(=O)C₂H₅	-H	5.47	8	1.18	1.31

[a] Not included in the derivation of Model 10.20.

deviation in log (% displaced) is 0.05, approximately that of replicates. For these same esters, the standard deviation in log (1/C_{SA}) is 0.49 against the Gram-positive organism, *Staphylococcus aureus*. On the other hand, the 2' esters bind less well to the ribosomes, but yet they are approximately as potent as the other esters because the 2' ester is easily hydrolyzed; that is, they act as pro-drugs to the corresponding

TABLE 10.11

Structure–Activity Models for Erythromycin Esters, Leucomycins, and Lincomycins

Compounds	Model	R^2	s	n	Model Number
Erythromycin esters	$\log (1/C_{SA}) = 2.98 - 0.303\pi$	0.72	0.26	22	10.20
Leucomycin	$\log (1/C_{BS}) = 2.56 + 0.416\pi$ $- 0.210\,S$	0.88	0.10	8	10.21
N-methyl and N-ethyl Lincomycins	$\log (1/C_{SL}) = -0.474 + 1.38\pi$ $- 0.245\pi^2 + 0.244\,T$	0.89	0.19	25	10.22
N-H Lincomycins	$\log (1/C_{SL}) = -0.080 + 0.342\pi$	0.75	0.11	7	10.23

2′-hydroxyl analog.[25] Hence, although binding to the ribosome is a prerequisite for antibacterial activity,[24] in the case of the alkyl esters of erythromycin, there is no correlation between the strength of binding to the target macromolecule and potency against bacteria.

In our first studies we used standard π values[26] to develop Model 10.20 (Table 10.11) for the 22 analogs for which potency against *S. aureus* was available at the time. It shows a negative dependence of potency on hydrophobicity. Table 10.10 shows that it has good predictive ability: In fact, it correctly extrapolates to predict the potency of the two most potent analogs. Because the model fits and predicts the antibacterial activity of both analogs that do and do not bind to the ribosome, we confirmed the observation that the relative antibacterial potency of these analogs is governed by penetration effects.

2. Effect of Hydrophobicity on the Antibacterial Activity of Leucomycin Esters

Leucomycin (Structure 10.6) is structurally similar to erythromycin, also inhibits bacterial protein synthesis, and was thought to bind to the same site on the bacterial ribosome as erythromycin.[27,28] Hence, to expand our knowledge of bacterial protein synthesis inhibitors, we used the literature reports of antibacterial potency to explore

(10.6)

TABLE 10.12

In Vitro Activity of Leucomycins against *B. subtilis*

				$Log (1/C_{BS})$	
R_1	R_3	π	S	Obs.	Calc. Model 10.21
-H	$-CH_2CH(CH_3)_2$	1.80	0.00	3.22	3.31
$-C(=O)CH_3$	$-CH_2CH(CH_3)_2$	1.80	1.00	3.09	3.10
-H	$-CH_2CH_2CH_3$	1.50	0.00	3.18	3.19
$-C(=O)CH_3$	$-CH_2CH_2CH_3$	1.50	1.00	3.00	2.98
-H	$-CH_2CH_3$	1.00	0.00	3.10	2.98
$-C(=O)CH_3$	$-CH_2CH_3$	1.00	1.00	2.89	2.77
-H	$-CH_3$	0.50	0.00	2.75	2.77
$-C(=O)CH_3$	$-CH_3$	0.50	1.00	2.43	2.56

Source: Martin, Y. C.; Lynn, K. R. *J. Med. Chem.* **1971**, 14, 1162–6.

the QSAR of the leucomycin antibiotics.[29] Table 10.12 lists the structure–activity relationships of antibacterial potency against *Bacillus subtilis*, another Gram-positive bacterium. For the QSAR, the indicator variable S was used to distinguish the 3-OH from the 3-O-acetyl analogs. Model 10.21 (Table 10.11) fits this data. Thus, the dependence of potency on hydrophobicity is positive for leucomycin esters—the opposite direction from that for the erythromycin esters.

3. Effect of Hydrophobicity on the Antibacterial Activity of Alkyl Analogs of Lincomycin

To further expand our knowledge of bacterial protein synthesis inhibitors, we also used literature data to explore the QSAR of analogs of lincomycin (Structure 10.7). It is another antibiotic thought to bind at the same ribosomal site as erythromycin and leucomycin.[30] Table 10.13 lists the structure–activity relationships against *Sarcina lutea*, another Gram-positive organism. The potency had been measured in a medium of approximately pH 7.8, using a 16–18 h agar diffusion assay.[31] Model 10.22 (Table 10.11) summarizes the QSAR for the *N*-alkyl analogs.[29] Note that the R_2 substituent can be either *cis* or *trans* to the amide link. Hence, we included an indicator variable to distinguish the molecules in which R_2 is *trans* to the amide ($T = 1.0$) from the *cis* ($T = 0.0$) analogs. In contrast to the QSARs for the erythromycins and

(10.7)

TABLE 10.13
Physical Properties and Potency of Lincomycin Analogs against *S. lutea*

R₁	R₂	π	Log P	pK$_a$	Log (1/C$_{SL}$) Obs.	Calc. Model 10.34
-H	-H	-0.5	-1.60	8.86[a]	-1.66[b]	-2.43
-H	-C₃H₇	1.0	0.02	8.86[a]	-1.31	-1.44
-H	-C₄H₉	1.5	0.56	8.86[a]	-2.00[b]	-1.12
-H	-C₅H₁₁	2.0	1.10	8.86[a]	-0.81	-0.79
-H	-C₆H₁₃	2.5	1.64	8.86[a]	-0.45	-0.48
-H	-C₇H₁₅	3.0	2.18	8.86[a]	-0.29	-0.22
-H	-C₈H₁₇	3.5	2.72	8.86[a]	-0.17	-0.10
-CH₃	-H	0.0	-1.06	7.65	-1.64[b]	-1.24
-CH₃	-C₂H₅ (*trans*)	1.0	0.02	7.65	-0.54	-0.36
-CH₃	-C₃H₇ (*trans*)	1.5	0.56	7.65	0.00	-0.04
-CH₃	-C₄H₉ (*trans*)	2.0	1.10	7.65	0.34	0.26
-CH₃	-C₅H₁₁ (*trans*)	2.5	1.64	7.65	0.56	0.48
-CH₃	-C₆H₁₃ (*trans*)	3.0	2.18	7.65	0.60	0.53
-CH₃	-C₇H₁₅ (*trans*)	3.5	2.72	7.65	0.23	0.36
-CH₃	-C₈H₁₇ (*trans*)	4.0	3.26	7.65	0.06	0.07
-CH₃	-C₃H₇ (*cis*)	1.5	0.56	7.65	-0.30	-0.26
-CH₃	-C₄H₉ (*cis*)	2.0	1.10	7.65	0.13	0.04
-CH₃	-C₅H₁₁ (*cis*)	2.5	1.64	7.65	0.28	0.26
-CH₃	-C₆H₁₃ (*cis*)	3.0	2.18	7.65	0.36	0.31
-CH₃	-C₇H₁₅ (*cis*)	3.5	2.72	7.65	0.01	0.14
-CH₃	-C₈H₁₇ (*cis*)	4.0	3.26	7.65	-0.16	-0.14
-C₂H₅	-H	0.5	-0.52	8.26[a]	-1.73[b]	-1.27
-C₂H₅	-C₃H₇ (*trans*)	2.0	1.10	8.26[a]	0.01	-0.08
-C₂H₅	-C₄H₉ (*trans*)	2.5	1.64	8.26[a]	0.10	0.20
-C₂H₅	-C₅H₁₁ (*trans*)	3.0	2.18	8.26[a]	0.52	0.36
-C₂H₅	-C₆H₁₃ (*trans*)	3.5	2.72	8.26[a]	0.35	0.30
-C₂H₅	-C₈H₁₇ (*trans*)	4.5	3.80	8.26[a]	-0.32	-0.25
-C₂H₅	-C₃H₇ (*cis*)	2.0	1.10	8.26[a]	-0.29	-0.30
-C₂H₅	-C₄H₉ (*cis*)	2.5	1.64	8.26[a]	-0.13	-0.02
-C₂H₅	-C₅H₁₁ (*cis*)	3.0	2.18	8.26[a]	0.15	0.14
-C₂H₅	-C₆H₁₃ (*cis*)	3.5	2.72	8.26[a]	0.13	0.08
-C₂H₅	-C₈H₁₇ (*cis*)	4.5	3.80	8.26[a]	-0.32	-0.47

Source: Martin, Y. C.; Lynn, K. R. *J. Med. Chem.* **1971**, 14, 1162–6.
[a] Results of the nonlinear fit.
[b] Not included in the fit of the data.

leucomycins, an optimum π is indicated. Perhaps because they have a different pK_a than the N-alkyl analogs, the N-H analogs (R_1 = H) fit a somewhat different QSAR, Model 10.23.

4. The Optimum Partition Coefficient for Erythromycin Esters

We now sought to estimate the partition coefficient that is associated with maximum potency in erythromycin esters. Is it lower than that of erythromycin A? If so, this would suggest that more hydrophilic esters would be more potent. Because the QSAR for leucomycin analogs showed that potency increases with increasing π but for erythromycin analogs it decreases with increasing π, putting both relationships on a log P scale would unite the two sets of observations. To do so, we needed measured values for erythromycin and leucomycin A1 (Structure 10.6, R_1 = H, R_3 = $CH_2CH(CH_3)_2$), the most potent analog. The erythromycin measurements are discussed in Chapter 9. We did not have a sample of leucomycin, so instead we used the measured log D of niddamycin, the structure of which differs from that of leucomycin in that the 9-OH is converted to a ketone. To the measured log P of niddamycin, 1.86, we added the difference in measured log D of 2-butanone and 2-butanol, 0.32, to give a value of 2.18 for the log P of leucomycin A1 and corrected for ionization at pH 8 to give a log P value of 2.88. Table 10.14 summarizes the optimum log P values for the three antibiotic classes. We concluded that the optimum log P of erythromycin analogs is not much lower than that of the log P of erythromycin A itself.

5. The Effect of Variation of the pK_a on the Potency of N-Benzyl Erythromycin Analogs

Another approach to increasing potency or expanding antibacterial spectrum of erythromycin would involve changing its ionization properties. In particular, if the pK_a's were lowered, a larger fraction of the analog would be present as the neutral form at the pH of testing, and this might facilitate penetration into the bacterial cell. The series of N-benzyl analogs listed in Table 10.15 was prepared to explore this point. Note that many of the analogs had been tested at more than one pH. This was done to explore the hypothesis that (1) the bacterium maintains an internal pH at 6.0 independent of the pH of the medium and (2) there is an equilibrium such that

---

TABLE 10.14

Comparison of Optimum Log P's for In Vitro Activity of Antibiotics that Bind to the 50S Subunit of the Ribosome

Antibiotic Class	Microorganism	Optimum Log P
Lincomycin	*Sarcina lutea*	1.1
Leucomycin	*Bacillus subtilis*	≥2.9
Erythromycin	*Staphylococcus aureus*	≤2.5

TABLE 10.15
Physical Properties and Potency of N-Benzyl Erythromycin Analogs

Substituent (Erythromycin A Derivative Unless Noted)	pK_a	pH_1	π	C_{in}/C_{out}	$E_{s,p}$	$E_{s,m}$	R_3	F_3	Log $(1/C_{SA})$ Obs.	Calc. Model 10.35
-H	7.98	6.00	0.00	1.00	1.24	1.24	0.00	0.00	1.36	1.46
-H	7.98	7.00	0.00	9.15	1.24	1.24	0.00	0.00	2.31	2.27
4-NO$_2$	6.77	6.00	−0.28	1.00	0.22	1.24	0.00	0.00	1.2	1.20
4-NO$_2$	6.77	7.00	−0.28	4.34	0.22	1.24	0.00	0.00	1.9	1.53
4-NH$_2$	9.00	7.00	−1.3	9.91	0.63	1.24	0.00	0.00	1.87	2.28
4-F	7.90	7.00	0.05	8.99	0.78	1.24	0.00	0.00	2.17	2.12
4-CN	6.94	6.00	−0.57	1.00	0.73	1.24	0.00	0.00	1.36	1.37
4-CN	6.94	7.00	−0.57	5.19	0.73	1.24	0.00	0.00	1.99	1.78
4-Cl	7.66	6.00	0.64	1.00	0.27	1.24	0.00	0.00	1.15	1.14
4-Cl	7.66	7.00	0.64	8.38	0.27	1.24	0.00	0.00	1.96	1.87
4-CH$_3$	8.15	6.00	0.46	1.00	0.00	1.24	0.00	0.00	1.09	1.08
4-CH$_3$	8.15	7.00	0.46	9.40	0.00	1.24	0.00	0.00	1.93	1.91
3-OCH$_3$	7.80	7.00	−0.02	8.77	1.24	0.69	−0.51	0.26	2.44	2.24
3-NO$_2$	6.88	6.00	−0.28	1.00	1.24	0.22	0.16	0.67	1.31	1.49
3-NO$_2$	6.88	7.00	−0.28	4.88	1.24	0.22	0.16	0.67	1.99	1.87
3-NH$_2$	8.23	6.00	−1.29	1.00	1.24	0.63	−0.68	0.02	1.66	1.57
3-NH$_2$	8.23	7.00	−1.29	9.50	1.24	0.63	−0.68	0.02	2.68	2.41
3-F	7.45	6.00	0.00	1.00	1.24	0.78	−0.34	0.43	1.38	1.46
3-F	7.45	7.00	0.00	7.64	1.24	0.78	−0.34	0.43	2.22	2.12
3-Cl	7.41	6.00	0.59	1.00	1.24	0.27	−0.15	0.41	1.34	1.41
3-Cl	7.41	7.00	0.59	7.48	1.24	0.27	−0.15	0.41	2.17	2.06
3-CH$_3$ (erythromycin B analog)	8.09	7.00	0.95	9.31	1.24	0.00	−0.13	−0.04	2.11	2.21
3-CH$_3$ (erythromycin B analog)	8.09	6.00	0.51	1.00	1.24	0.00	−0.13	−0.04	1.38	1.42
3-CH$_3$ (erythromycin B analog)	8.09	7.00	0.51	9.32	1.24	0.00	−0.13	−0.04	2.38	2.25
3,5-CH$_3$	8.19	7.00	0.73	9.45	1.24	0.00	−0.26	−0.08	2.36	2.24
2-F	7.30	7.00	0.14	7.00	1.24	1.24	0.00	0.00	2.16	2.05
2-CH$_3$ (erythromycin B analog)	8.15	7.00	0.52	9.40	1.24	1.24	0.00	0.00	1.96	2.25
2-CH$_3$ (erythromycin B analog)	8.15	7.00	0.08	9.40	1.24	1.24	0.00	0.00	2.21	2.29
2-CH(CH$_3$)$_2$,5-NO$_2$	7.38	7.00	1.2	7.35	1.24	1.24	0.16	0.67	1.59	1.99
2,5-CH$_3$	8.30	7.00	0.59	9.57	1.24	1.24	−0.13	−0.04	2.14	2.26
2,4,6-CH$_3$	8.30	7.00	0.67	9.57	0.00	1.24	0.00	0.00	1.69	1.91
2,3-(-C$_4$H$_4$-)	7.98	7.00	1.46	9.15	1.24	0.00	0.01	0.03	2.34	2.15
3-CH$_3$	8.09	7.65[a]	0.51	33.04	1.24	0.00	−0.13	−0.04	2.51	4.59

(Continued)

TABLE 10.15 (CONTINUED)
Physical Properties and Potency of *N*-Benzyl Erythromycin Analogs

Substituent (Erythromycin A Derivative Unless Noted)	pK_a	pH_1	π	C_{in}/C_{out}	$E_{s,p}$	$E_{s,m}$	R_3	F_3	Log (1/C_{SA}) Obs.	Calc. Model 10.35
3-F	7.45	7.65[a]	0.00	17.89	1.24	0.78	−0.34	0.43	2.29	3.13
3-NH$_2$	8.23	7.65[a]	−1.29	35.57	1.24	0.63	−0.68	0.02	2.93	4.99
4-CH$_3$	8.15	7.65[a]	0.46	34.18	0.00	1.24	0.00	0.00	2.14	4.36
4-Cl	7.66	7.65[a]	0.64	23.09	0.27	1.24	0.00	0.00	2.02	3.32
4-CN	6.94	7.65[a]	−0.57	8.13	0.73	1.24	0.00	0.00	1.89	2.07

[a] Not included in the fit of the model.

the concentration of the uncharged form is equal in the medium and inside the cell. From these assumptions, the definition of pK_a, and material balance, the following relationship may be derived:

$$\frac{[C_{in,i}]}{[C_{out,i}]} = \frac{K_{a,i} + [H^+]_{in}}{K_{a,i} + [H^+]_{out}} \tag{10.24}$$

The pK_a's of a few analogs had been measured. The σ values of the substituents of these analogs established a ρ value that in turn was used to calculate the pK_a's of the remaining molecules. The table also lists the hydrophobic π values and $E_{s,p}$, $E_{s,m}$, R_3, and F_3. The π values were calculated from the log P values of substituted toluenes.[32] Model 10.25 is the best fit of the observations at pH's 6.00 and 7.00:

$$\log(1/C_{SA}) = 1.015 - 0.084\pi + 0.099([C_{in}]/[C_{out}]) + 0.281E_{s,p} \tag{10.25}$$

$$R^2 = 0.926, s = 0.118, n = 32$$

(We omitted the observations at pH 7.85 because of these results were less reproducible than those at other pH's.) Model 10.25 supports the initial assumption that differences in the pK_a of the molecule affect the amount of an analog that partitions into the bacterium. However, pH-partitioning is not the complete answer because the slope of Model 10.25 is ten times lower than the theoretical value of 1.0, and the multiparameter model is a substantial improvement over that with only $[C_{in}]/[C_{out}]$ for which R^2 and s are 0.720 and 0.234. Thus, in addition to the effect of pK_a on potency, there appears to be a positive steric effect of substituents and a negative hydrophobic effect of all substituents.

B. FURTHER STUDIES ON ALKYL ESTERS (1971)

Chapter 9 details these studies, which are based on measured partition coefficients. The 2D QSAR models suggest that it is possible that decreases in log P could lead to an increase in potency, but indicator variables suggest a possible negative steric effect of substituents.

C.   POTENCY OF NEW ANALOGS AGAINST *S. AUREUS* (1972–1973)

As the synthetic methodologies for handling erythromycin were expanded, it became possible for the medicinal chemists to prepare analogs in which the ester group contains an oxygen or nitrogen atom (Table 10.16). Including these analogs in the QSAR breaks the previous correlation between hydrophobicity and size, and it also extends the range of hydrophobicity to values lower than that of erythromycin.

The predicted potencies of these analogs are also listed in the table. Clearly, Models 9.3 and 9.5 overpredict the potency of the less hydrophobic analogs. This suggested that the new analogs might have identified the log *P* associated with optimum potency.

TABLE 10.16
Observed and Calculated Potencies of Newer Carboxylate Esters of Erythromycin

Analog	Log *P*	*A*	D4	Log (1/C_{SA})			
				Obs.	Pred. Model 9.3[a]	Pred. Model 9.5[b]	Calc. Model 10.26
Included in the Derivation of Model 10.26							
4″-*O*-glycyl A	1.65[c]	1.00	1.00	2.31	3.81	4.00	2.10
4″-*O*-glycyl B	2.09[c]	0.00	1.00	2.15	3.32	3.76	2.25
4″-*O*-methoxyacetyl B	2.87	0.00	1.00	2.01	2.45	2.70	2.27
4″-*O*-methylsuccinyl B	3.24	0.00	1.00	2.30	2.03	2.19	2.19
4″-*O*-carbomethoxy B	3.25	0.00	1.00	2.19	2.02	2.18	2.19
Not Included in the Derivation of Model 10.26							
4″-*O*-(*N*-acetyl)glycyl A	1.55[c]	1.00	1.00	2.09	3.92	4.13	2.06
2′-*O*-acetyl-4″-*O*-glycyl A	1.88[c]	1.00	1.00	2.18	3.55	3.68	2.19
4″-*O*-glycolyl B	2.09[c]	0.00	1.00	2.18	3.32	3.76	2.25
4″-*O*-carbobenzoxyglycyl A	2.81	1.00	1.00	2.82	2.51	2.42	2.19
2′-*O*-acetyl-4"-*O*-carbobenzoxyglycyl A	3.04	1.00	1.00	2.59	2.26	2.11	2.14
2′-*O*-acetyl-4"-*O*-carbomethoxy B	3.10	0.00	1.00	2.12	2.19	2.38	2.18
4″-*O*-trimethylacetyl B	3.43	0.00	1.00	2.24	1.82	1.94	2.01
4″-*O*-benzyloxyacetyl B	3.93	0.00	1.00	2.59	1.26	1.26	1.65

[a] Model 9.3: $\log(1/C_{SA}) = 6.01 - 1.12\log P - 0.35 D4$ $\qquad R^2 = 0.889, s = 0.174$
[b] Model 9.5: $\log(1/C_{SA}) = 6.89 - 1.36\log P - 0.29 D4 - 0.36A - 0.17 D11$ $\quad R^2 = 0.945, s = 0.127$
[c] Log *P* is outside the range of Models 9.3 and 9.5.

Because the log P of each of the remaining-analogues is within the range of those studied in Chapter 9, their predicted potencies are not based on extrapolation and hence should be reliable. Model 9.3, which is based on log P and is an indicator for 4″-substitution, provides more accurate predictions than does Model 9.5, which has an additional indicator for erythromycin A derivatives. However, the prediction errors are much larger than those from the fit and, in general, potency is underestimated. Because the substituents of the new analogs are larger than those training set, the larger error suggest that a QSAR model with a better description of steric effects should be sought.

To expand the scope of the QSAR model to these more varied analogs, we needed better estimates of log P and steric effects. Accordingly, we measured the log P of additional analogs to provide the information needed to estimate the log P of those remaining. We estimated the steric properties listed in Table 10.17 from the $E_{s,c}$ value of the corresponding alkyl group. The resulting QSAR, Model 10.26 (Table 10.18), suggests that 2.47 is the log P that corresponds to optimum antibacterial potency.

Table 10.16 lists the fitted or predicted potencies. The potency was correctly predicted for four of the eight analogs that had not been included in the fit because their potencies were not known at the time. The outliers are all more potent than predicted and are the only analogs that contain a benzyloxy group.

We tested the predictivity of Model 10.26 with the analogs listed in Table 10.19. This set includes esters related to those included in the models, which we expected to be accurately predicted. However, it also includes molecules different from the training set: sulfonate esters, derivatives of a hydroxyl added to position 4′, and analogs substituted at the ketone (position 9). Figure 10.5 shows that except for one or

TABLE 10.17
$E_{s,c}$ Values Used in the Calculations

Group	Structure	$E_{s,c}$
Hydroxyl	-OH	0.00
Formyl	-OC(=O)H	−0.33
Acetyl	-OC(=O)CH$_3$	−0.67
Propionyl	-OC(=O)C$_2$H$_5$	−0.69
Glycyl	-OC(=O)CH$_2$NH$_2$	−0.64
Methoxy acetyl	-OC(=O)CH$_2$OCH$_3$	−0.64
Carbomethoxy	-OC(=O)OCH$_3$	−0.64
Methylsuccinyl	-OC(=O)(CH$_2$)$_2$C(=O)OCH$_3$	−0.64
Glycolyl	-OC(=O)CH$_2$OH	−0.64
Trimethylacetyl	-OC(=O)C(CH$_3$)$_3$	−0.76
p-nitrobenzyl	-OC(=O)C$_6$H$_4$-4-NO$_2$	−0.76
Carbobenzoxyglycyl	-OC(=O)CH$_2$NHC(=O)OCH$_2$C$_6$H$_5$	−0.76
Benzyloxyacetyl	-OC(=O)CH$_2$OCH$_2$C$_6$H$_5$	−0.76
Methylthiomethoxy	-OCH$_2$SCH$_3$	−0.69

TABLE 10.18
New Property-Based 2D QSAR Models for Erythromycin Analogs

				Data Considered				
Model	10.26	10.27	10.28	10.29	10.30	10.31	10.32	10.33
Organism	S. aureus	S. aureus	S. aureus	S. aureus	K. pneumoniae	K. pneumoniae	E. coli	H. influenzae
Gram	Positive	Positive	Positive	Positive	Negative	Negative	Negative	Negative
Type of assay	Turbid.	Turbid.	Turbid.	2× Dilution	Turbid.	2× Dilution	2× Dilution	2× Dilution
Analogs	Esters	All[a]	Meas. log P	All[a]	All[a]	All[a]	All[a]	All[a]
n	33	65	22	43	35	37	32	37
			Coefficients of Regression Equation					
Intercept	1.26	1.10	1.00	-1.83	1.37	-1.02	-2.70	-3.38
Log P	1.33	1.54	1.60	1.58	1.46	1.08	2.56	2.79
(Log P)^2	-0.27	-0.32	-0.34	-0.35	-0.32	-0.26	-0.56	-0.56
$E_{s,4''}$	0.89	0.94	0.73	0.80	0.38	0.62	0.65	0.83
$E_{s,11}$	0.78	0.62	—	0.38	0.48	—	—	1.71
S	—	0.19	—	—	—	—	—	—
			Statistics					
R^2	0.93	0.93	0.76	0.77	0.9	0.64	0.66	0.85
s	0.13	0.11	0.23	0.22	0.12	0.24	0.32	0.34
Optimum log P	2.47	2.39	2.39	2.27	2.25	2.1	1.97	2.46

[a] Exceptions noted in text.

TABLE 10.19

Observed and Predicted Potencies of Analogs of Erythromycin not Included in the Establishment of Model 10.26

Analog	Erythromycin A or B	Log P	Log $(1/C_{SA})$ Obs.	Predicted Model 10.26
Group I—Sulfonate Esters				
2'-O-acetyl-4"-O-methanesulfonoxy	B	3.39	2.13	2.08
4"-O-methanesulfonoxy	B	3.16	2.31	2.18
11-O-methanesulfonoxy	B	3.16	2.56	2.18
4"-O-methanesulfonoxy	A	2.72	2.64	2.29
Group II—Other 2', 4", or 11-Substituted Analogs				
2'-O-benzoyl	A	3.87	2.14	2.18
2'-O-propionyl	A	3.04	2.65	2.79
2'-O-ethylsuccinyl	A	2.77	2.89	2.88
Erythromycin	C	1.25[a]	2.47	2.61
2'-O-formyl	B	3.02	2.78	2.80
11-methylthiomethoxy	B	3.13	2.38	2.34
2'-O-succinyl	B[b]	2.83	2.18	2.18
Group III—4'-Hydroxy Analogs				
4'-Hydroxy	A	1.44	2.72	2.77
2'-Acetyl-4'-acetoxy	A	1.90	2.91	2.73
2',4'-Dibenzoxy	A[b]	3.56	0.46	2.46
Group IV—Analogs Substituted on the Ketone				
Hydrazone	A	2.98	2.60	2.82
Hydrazone	B	3.42	2.59	2.56
Oxime	B	3.42	2.56	2.63
Oxime	A	2.98[a]	2.81	2.81
9-O-methyl oxime	B	3.71	2.33	2.53
9-O-methyl oxime	A	3.27[a]	2.66	2.73
9-O-ethyl oxime	B	3.50[b]	2.63	2.50
9-O-ethyl oxime	A	3.50	2.84	2.50
9,11-Carbonate-6,9-hemiketal	A	2.56[a]	3.09	2.94
9,11-Phenylboronate-6,9-hemiketal	A	3.09[a]	2.96	2.77
9,11-Butylboronate-6,9-hemiketal	A	2.93	2.83	2.89
9,11-Cyclohexylboronate-6,9-hemiketal	A	3.19	2.72	2.71
9,11-Methylboronate-6,9-hemiketal	A	2.24	2.91	2.96
Group V—Esters of the 9-oxime of Erythromycin B				
Benzoyl	B	3.51	2.63	2.50
2',9-Dibenzoyl	B	4.57	1.17	1.33
Trimethylacetyl	B	3.51	2.45	2.50

(Continued)

TABLE 10.19 (CONTINUED)
Observed and Predicted Potencies of Analogs of Erythromycin not
Included in the Establishment of Model 10.26

| | | | Log (1/C_{SA}) | |
Analog	Erythromycin A or B	Log P	Obs.	Predicted Model 10.26
Mesitoyl	B	3.51[a]	2.45	2.50
(2,5-Dimethyl-4-carbomethoxy)-benzoyl	B	3.44[a]	2.68	2.55
(2,6-Dimethyl-4-N-butylcarbamyl)-benzoyl	B	3.48	2.19	2.52
(2,6-Dimethyl-4-morpholinylcarbonyl)-benzoyl	B	2.44	2.27	2.96
Mesitoyl ester	A	3.07	2.76	2.77
Group VI—Miscellaneous				
Niddamycin		2.00	3.05	2.93
8,9-Anhydro-6,9-hemiketal B		4.62[a]	1.16	1.26

[a] Measured value.
[b] There is probably an error in this calculation, but this is the value that was used in the derivation of
     the model.

two analogs, Model 10.26 provides accurate predictions. The largest outlier is the
(2,6-dimethyl-4-morpholinylcarbonyl)-benzoyl ester of erythromycin B oxime and
the other is 2′,4′-dibenzoxy erythromycin A.

Model 10.27 (Table 10.18) describes the QSAR for 65 analogs. The data used for
fitting the model omitted three analogs whose structure is very different from others

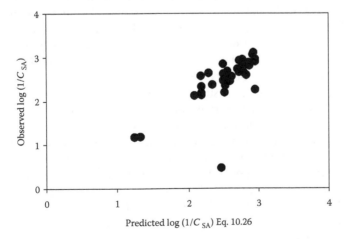

FIGURE 10.5   For the erythromycin analogs in Table 10.19, a plot of the observed potency
versus that predicted from Model 10.26.

in the data set: 2'-O-succinylerythromycin B, 2',4'-dibenzoxyerythromycin A, and the 8,9-anhydro-6,9-hemiketal of erythromycin B. The QSAR based on 65 analogs confirms that optimum activity is seen at a log P of 2.4 and that there are negative steric effects of substituents at the 4″ or 11 positions. It also suggests that, considering all factors, sulfonate esters are slightly more potent than predicted. The model is interesting in that steric effects are not needed for analogs substituted at the ketone.

Because some of the log P values used to establish Model 10.27 are calculated, not measured, we derived Model 10.28 using only the 22 analogs for which log P had been measured. It confirms the optimum log P of 2.4 and the negative steric effect of 4″ substituents. Compared to Model 10.27, two terms are missing: The steric effect of substituents at position 11 was not statistically significant, and the indicator S is not included because there were no sulfonyl esters in the sample. It is not clear why Model 10.28 has a larger s.

In summary, we expect Model 10.27 to accurately predict the potency of new analogs because it fits the potency of 65 analogs that are substituted at any of seven positions. However, such potency predictions are limited to analogs that bind to the ribosome, something that we could not predict.

D. POTENCY OF ALL ANALOGS AGAINST OTHER BACTERIA (1971–1973)

In order to develop a QSAR for activity against Gram-negative organisms, we needed relative potency values against such organisms. This presented a problem because the turbidometric method that is used to measure potency of versus *S. aureus* could not be used for any clinically significant Gram-negative organism, only for *Klebsiella pneumoniae*. However, we did have minimum inhibitory concentrations against Gram-negative organisms in twofold agar dilution assays, a more qualitative indication of relative potency. Table 10.18 compares QSAR models developed with potencies measured against *S. aureus* and *K. pneumoniae* with turbidometric methods, Models 10.26 and 10.30, with those measured with agar dilution assays, Models 10.29 and 10.31. Although the latter are less precise, the same relationships are apparent except that the optimum log P values are slightly lower for the agar dilution method. We concluded that reliable QSARs could be derived from potencies measured with the tube dilution method.

In contrast to the literature studies on antibiotics with a different mode of action,[21,22] the QSARs on erythromycin analogs suggest that the optimum log P is the same for potency against both Gram-positive and Gram-negative bacteria. Thus, Model 10.32 for activity against the Gram-negative bacterium *Escherichia coli* suggests an optimum log P only approximately 0.3 log units lower than that for Model 10.29 for the Gram-positive *S. aureus*, but Model 10.33 for activity against another Gram-negative bacterium, *Haemophilis influenzae*, suggests an optimum log P 0.19 higher than Model 10.29. Thus, although we knew that the newer analogs did not show enhanced Gram-negative activity, the models show that this is not due to suboptimal log P.

The QSARs do not suggest what properties could be modified to increase Gram-negative potency. In fact, they suggest that potency against Gram-negative organisms should be approximately the same as against Gram-positive organisms—ignoring steric effects; the relative log $(1/C_{SA})$ at the optimum log P is very similar for the three organisms. From Model 10.29, the calculated log $1/C$ at the optimum log P for *S. aureus* is -0.047; from Model 10.30, it is 0.095 for *H. influenzae*; and from Model 10.33, it is 0.170 for *E. coli*, corresponding to minimum inhibitory concentrations of 0.90, 1.24, and 1.48 µg/mL. However, the possible importance of steric effects is suggested by the large positive coefficient of E_s (a strong negative influence of larger substituents) in Model 10.33 for molecules substituted at position 11.

E.   OVERVIEW OF CONCLUSIONS FROM 2D QSARs (1973) OF ERYTHROMYCIN ANALOGS

- The log P associated with optimum inhibition of growth of both Gram-positive and Gram-negative organisms is approximately that of erythromycin A.
- Once the optimum log P was established, the models predicted the potency of diverse molecules with a standard deviation of the error of 0.52.
- Steric effects decrease the potency of esters of the 2', 4", and 11-hydroxyls as well as para-substituted *N*-benzyl analogs.

F.   FITS TO MODEL-BASED EQUATIONS (1975)

1.   Lincomycin Analogs[33]

Although Models 10.22 and 10.23 (Table 10.11) fitted the structure-antibacterial potency of lincomycin analogs, we investigated the possibility that fits to model-based equations would provide additional insights. The model of interest is shown in Scheme 10.1. Table 10.13 lists the data used. The pK_a of lincomycin[34] and the log P values[31] for lincomycin and *N*-demethyl-lincomycin are literature values. We used the measured pK_a of lincomycin for all of the analogs in which R_1 is $-CH_3$ and fitted the pK_a's for the analogs in which R_1 is $-C_2H_5$ and $-H$. The log P values of all other analogs were calculated.[35] We omitted the four analogs that have a

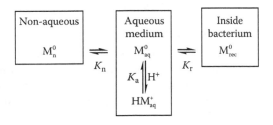

SCHEME 10.1

minimum inhibitory concentration in the order of 400 µg/mL, log $(1/C_{SL})$ less than −1.5, because potency this low is very difficult to quantitate.

Because lincomycin equilibrates between the various compartments within less than 1 h in a broth dilution assay,[36] we assumed that the compounds would equilibrate in the 16–18 h agar diffusion assay that was used for these compounds. Hence, we used equilibrium models to fit the data.

A good fit is found for the simple model with aqueous medium, nonaqueous, and internal bacterial compartments. However, the fit is improved when an indicator variable equal to 1.0 for *trans* analogs is used, and when the pK_a's of the analogs for which R_1 is −*N*-H or −*N*-Et are also fitted (Model 10.34):

$$\log(1/C_{SL}) = \log \frac{1}{1 + P^c + \dfrac{a(1 + 10^{pK_a - pH})}{P^b}} + D_{trans} + X \qquad (10.34)$$

$$R^2 = 0.958, \quad s = 0.099, \quad n = 28$$

$$\log a = 2.22 \pm 0.24, \quad b = 0.61 \pm 0.06, \quad c = 0.61 \pm 0.08, \quad pK_{a,NEt} = 8.26 \pm 0.10,$$
$$pK_{a,NH} = 8.86 \pm 0.08, \quad D_{trans} = 0.22 \pm 0.04, \quad X = 1.86 \pm 0.24$$

Because of the excellent fit to Model 10.34, we concluded that there are hydrophobic interactions of the analogs to both a nonaqueous phase and to the target biomolecule and that *trans* analogs are somewhat more potent than the corresponding *cis* analogs. The stereochemistry of the R_2 substituent was not reported for the NH analogs; however, they fit the models with $D_{trans} = 0$.

The model also suggests that only the neutral form binds to the target. Although the structure–activity data do not justify including the additional terms required, the true situation could be an equilibrium of two aqueous compartments of different pH and binding to the target of the charged form with affinity proportional to the pK_a.

The fitted pK_a of the *N*-Et analogs is in the appropriate direction and relative order from lincomycin, $pK_a = 7.6$. One reported pK_a of an *N*-Et derivative is 8.0, the other 8.2. On the other hand, the fitted pK_a of the *N*-H analogs deviates more from the experimental value of 8.4.[19] The discrepancy could be due to uncertainty in the experimental measures of potency, of log P, of pK_a, or to a lack of correspondence between the model and the reality.

Compared to the fits to the 2D models, Models 10.21 and 10.22 in Table 10.11, we see that fits to the model-based model has a slightly smaller standard error of estimate and that one model fits all of the data.

2. *N*-Benzyl Erythromycin Analogs

The data are listed in Table 10.15, and the model is shown in Scheme 10.2. The data are well fitted by a model that has two aqueous compartments of different pH with pH_2 fixed at 6.0, binding of only the protonated form of the molecule to the ribosomal RNA, electronic effects of the meta substituent, and a negative steric effect of the para substituent, Model 10.35:[33]

$$\log(1/C) = \log \frac{1 + 10^{pK_a - pH_2}}{1 + 10^{pK_a - pH_2} + a(1 + 10^{pK_a - pH_1})} + eE_{s,4} + fF_3 + rR_3 + X \quad (10.35)$$

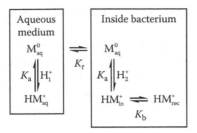

SCHEME 10.2

$$R^2 = 0.923, \quad s = 0.132, \quad n = 38$$

$$\text{Log } a = 1.13 \pm 0.11, \quad e = 0.25 \pm 0.056, \quad f = -0.23 \pm 0.11,$$
$$r = -0.55 \pm 0.11, \quad X = 2.37 \pm 0.15$$

From Model 10.35, we conclude that two aqueous phases of different pH are involved in the relative potency of the analogs: The pH inside the bacterium remains constant as the pH of the external phase is varied. The model also suggests that the substituents do not participate in hydrophobic interactions but that steric and electronic factors are present. These conclusions are different from those from Model 10.25, which suggested a modest influence of hydrophobicity.

Although Model 10.35 suggests that the positively charged species is biologically active, the relative potency of the ionic and neutral forms cannot be assessed nor can the apparent pH inside the bacterium be established. This is because in Equation 6.34, which results from the more elaborate model that would be required, there is a canceling of terms because of two features of the data set: First, because the pH inside the cell is approximately 6.0, below the pK_a's of the compounds, the concentration of the ionic form inside the cell is 6–140 times higher than that of the neutral form. Second, because of the small volume of the bacterial cells compared to that of the external medium, most of the compound added to the assay remains in the external aqueous phase. Hence, there is not enough information to fit the relative potencies of the neutral and charged forms and also the pH inside the bacterium.

The term with a positive coefficient for E_s indicates steric inhibition of activity; the F and R terms for substituents at the meta position are less straightforward to interpret.

Several years ago, it was suggested that the relative potency of these analogs should be proportional to the ratio of concentration of analog inside the bacterium to that outside. In model form this relationship is

$$\log(1/C_{SA}) = \log \frac{10^{pK_a - pH_2}}{10^{pK_a - pH_1}} + X \tag{10.36}$$

$$R^2 = 0.682, \, s = 0.253, \, n = 38$$

Comparison of Model 10.36 with Model 10.35 shows that it is just a special case of the latter.

The studies quoted above confirm the expectation that fits to model-based equations are more interpretable than fits to empirical equations. A second advantage is that experiments conducted at several pH's may be included in one analysis. Although these models are harder to fit than linear ones, the increase in difficulty is not troublesome.

G. Free–Wilson Analysis to Complement Property-Based QSAR

A Free–Wilson analysis, because it ignores physical properties but simply considers the contribution of each substituent to potency, provides alternative insights into these structure–activity relationships. Table 10.20 lists the data for the Free–Wilson analysis of the original set of alkyl esters.[37] Note that separate descriptors are used for each position even if the same substituent is used at more than one position. Also note that, for each position one fewer descriptor than the number of substituents is needed—we use erythromycin A (Structure 1.9) as the reference so that descriptors are not needed if the 2′, 4″, or 11 position is not esterified.

FO2 and FO11 are not significantly different from zero. This agrees with the QSAR Models 9.1–9.5 in that this structural change does not change log P, but it contradicts Model 9.5, which suggests that esterification of the 11-hydroxyl decreases log $(1/C_{SA})$ by 0.17 log units.

The effect of the statistically significant contributions of substituents to potency is summarized in Model 10.37:[37]

$$\log(1/C_{SA}) = 2.75 - 0.15AC2 - 0.28FO4 - 0.68AC4 - 0.73PR4 \quad (10.37)$$
$$-0.54AC11 - 0.67PR11 + 0.23A$$
$$R^2 = 0.98, \quad s = 0.07, \quad n = 28$$

The R^2 and s of the Free–Wilson model show that it fits the data more precisely than does the best QSAR model of these analogs, Model 9.5. Although this might warn one that the Free–Wilson model overfits log $(1/C_{SA})$, it might also suggest that the physical properties used in the QSAR are somehow deficient: Either they are calculated incorrectly or another property is also involved. An example of the disparity between the two approaches is that, although the log P calculations used the same π values for the same substituent in the three positions of substitution, in the Free–Wilson model the coefficient of AC2 is very different from those of AC4 and AC11, which are not identical. A further discrepancy is that in the Free–Wilson model the coefficient of A is positive, but it is negative in Model 9.5. Taken together, these observations suggest that either the log P calculations are in error or that the QSAR models have not captured the more subtle relationships between potency and physical properties.

To investigate the possible improvement of the QSAR model, we explored log P values calculated with CLOGP[13] and KowWin.[5] Models 10.38 and 10.39 are the best we could find:

TABLE 10.20
Data for the Free–Wilson Analysis of Erythromycin Esters

| | | | | | | | | | Log $(1/C_{SA})$ | |
| | | | | | | | | | | Fitted Model |
Analog	FO2	AC2	FO4	AC4	PR4	FO11	AC11	PR11	A	Obs.	10.37
1.09	0	0	0	0	0	0	0	0	1	3.00	2.98
1.11	1	0	0	0	0	0	0	0	1	2.91	2.98
1.10	0	0	0	0	0	0	0	0	0	2.82	2.75
1.12	1	0	0	0	0	0	0	0	0	2.78	2.75
1.13	0	1	0	0	0	0	0	0	1	2.75	2.83
1.14	0	0	1	0	0	0	0	0	1	2.72	2.70
1.15	1	0	1	0	0	0	0	0	1	2.71	2.70
1.16	0	0	0	0	0	1	0	0	0	2.71	2.75
1.17	0	1	1	0	0	0	0	0	1	2.70	2.55
1.18	0	1	0	0	0	0	0	0	0	2.54	2.60
1.19	0	0	1	0	0	0	0	0	0	2.54	2.47
1.20	1	0	1	0	0	0	0	0	0	2.50	2.47
1.21	0	0	1	0	0	1	0	0	0	2.42	2.47
1.22	1	0	1	0	0	1	0	0	0	2.39	2.47
1.23	0	0	0	0	0	0	1	0	0	2.29	2.21
1.24	0	1	1	0	0	0	0	0	0	2.28	2.32
1.25	0	0	0	1	0	0	0	0	1	2.24	2.30
1.26	0	1	0	1	0	0	0	0	1	2.15	2.15
1.27	0	1	0	0	0	0	1	0	0	2.11	2.06
1.28	0	1	0	1	0	0	0	0	0	2.04	1.92
1.29	0	0	0	1	0	0	0	0	0	2.02	2.07
1.30	0	0	0	0	1	0	0	0	0	2.02	2.02
1.31	0	1	0	0	1	0	0	0	0	1.87	1.87
1.32	0	1	1	0	0	0	1	0	0	1.65	1.78
1.33	0	0	0	1	0	0	1	0	0	1.51	1.53
1.34	0	1	0	1	0	0	1	0	0	1.39	1.38
1.35	0	0	0	0	1	0	0	1	0	1.38	1.35
1.36	0	1	0	0	1	0	0	1	0	1.18	1.21

$$\log(1/C_{SA}) = 3.72 - 0.45 \times CLOGP - 0.15 \times D4 \tag{10.38}$$
$$R^2 = 0.90, \quad s = 0.16, \quad n = 28$$

$$\log(1/C_{SA}) = 3.30 - 0.28 \times KOWWIN - 0.36 \times D4 \tag{10.39}$$
$$R^2 = 0.64, \quad s = 0.32, \quad n = 28$$

Neither model fits the data as well as the Free–Wilson model.

The most serious limitation of the Free–Wilson model is that it cannot be used to predict the potency of any molecule that contains a substituent that is not included in its data matrix. Hence, the best use of this approach is to gain an impression of the statistics of maximum fit of a data set.

H.   COMFA 3D QSAR ANALYSIS TO COMPLEMENT PROPERTY-BASED QSAR

We explored whether a CoMFA 3D QSAR[15] analysis might be more predictive than the 2D QSAR based on the hydrophobic effects of molecules (Model 9.5).

To generate the 3D structures of the erythromycin esters, we assumed that they all act in the same conformation and that they can be superimposed over the common parts of this conformation. Thus, we first made a generic template of the most highly substituted molecule from the x-ray structure of 6-O-methylerythromycin A.[38] To do so, we removed the added 6-methyl group, converted the 2'-hydroxyl to O-acetyl and the 4''- and 11-hydroxyls to O-propionyl, and eliminated close contacts by minimizing the energy with the Sybyl force field.[39] We modified this template to prepare the 3D structure of each analog, adjusting the positions of any proton on an oxygen that is not esterified to that of the original x-ray structure. The steric lattice interaction energies of these 3D structures were calculated with the CoMFA 3D QSAR[15] defaults: 2 Å spacing and energy truncation at 30 kcal/mol.

The CoMFA 3D QSAR analysis of these structure–activity relationships shown in Table 10.21 attributes the changes in potency to changes in shape properties. Models with one, two, or six latent variables are supported by the large decrease in cross-validated s and increase in q^2. The superiority of CoMFA 3D QSAR fits to those from 2D QSAR underscores the possibility that the structure–activity relationships are better described as steric rather than hydrophobic. A complication in this conclusion is that steric properties and log P are not independent. Rather, for this series of esters, PLS using CoMFA 3D QSAR steric fields to predict the log P calculated from measured values requires six latent variables but results in an R^2 of 0.996 and q^2 of 0.987.

TABLE 10.21
Statistics of the CoMFA Analysis of the Potency of Erythromycin Esters ($n = 28$)

	Fitted Log($1/C_{SA}$)		Cross-Validated Log ($1/C_{SA}$)		
Latent Variables	s	R^2	s	q^2	Model
0	0.503	0.000	0.503	0.000	
1	0.160	0.903	0.192	0.860	10.40
2	0.129	0.939	0.153	0.914	10.41
3	0.121	0.948	0.150	0.921	10.42
4	0.109	0.960	0.140	0.934	10.43
5	0.098	0.969	0.134	0.942	10.44
6	0.069	0.985	0.108	0.964	10.45

The statistics for the CoMFA 3D QSAR fit with six latent variables is identical to that for the Free–Wilson model, which also contains six variables. This supports the idea that the Free–Wilson model describes steric effects of substituents.

Figure 10.6 shows the regions in space occupied by negative CoMFA 3D QSAR steric coefficients. Note that the regions grow as the number of latent variables is increased and s decreases. The negative effects of substitution at the 2'-position are seen best when six latent variables are included.

Table 10.22 shows that the least predictive model is the 2D QSAR based on 28 analogs (Model 9.5), with a standard deviation of prediction of 0.91. Using these same analogs, the CoMFA 3D QSAR models are more predictive, with standard deviations of 0.51, 0.53, and 0.56 for the 1, 2, and 6-latent variable models, respectively. The decline in predictivity suggests that adding more latent variables has overfitted the data even though the q^2 increases with more latent variables.

However, the most predictive model is Model 10.26, which established the optimum log P by adding only six molecules to the original data set of 28, and which decreases the prediction errors by more than half. Even the best CoMFA 3D QSAR model is not quite as predictive as the best 2D QSAR model.

I. LESSONS LEARNED FROM THE ANALYSIS OF STRUCTURE–ACTIVITY RELATIONSHIPS OF ERYTHROMYCIN ANALOGS

1. Property-based 2D QSAR analysis suggests that the log P that corresponds to optimum antibacterial potency of erythromycin analogs is the same for Gram-negative as for Gram-positive organisms. It provides no clues as to how changes in physical properties might change the antibacterial spectrum.

2. Although QSARs based on alkyl esters correctly predicted the potency of analogs within the log P range of the original model, they failed for more polar molecules. Adding just five polar molecules to the original 28 alkyl analogs produced the most predictive 2D QSAR.

3. Free–Wilson and CoMFA (steric fields only) QSAR analysis of the alkyl ester data set yielded models that fitted the data better than the property-based 2D QSAR. However, the CoMFA 3D QSAR models were slightly less predictive than those from property-based 2D QSAR.

4. Because it is not possible to unambiguously decide if hydrophobicity, as suggested by QSAR, or size and shape, as suggested by CoMFA 3D QSAR, are the primary determinants of potency, the set of analogs was not appropriately designed for these analyses.

5. Fits to model-based models suggest that the positively charged form of the analogs is biologically active, but that the uncharged form may be active also.

6. The results emphasize the importance of considering any computational model to be a hypothesis and of considering multiple hypotheses to

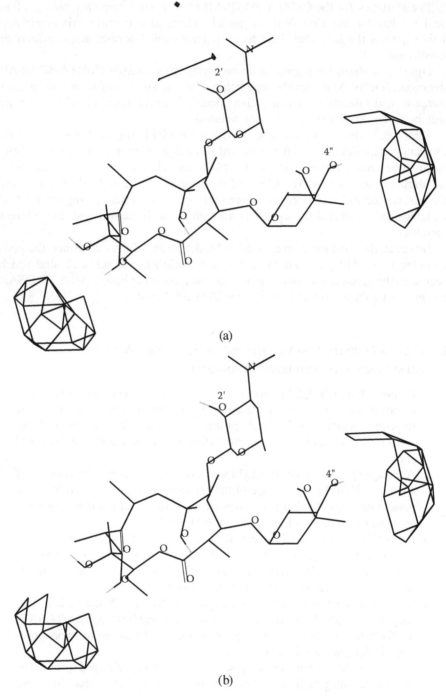

(a)

(b)

FIGURE 10.6 The regions in space occupied by negative CoMFA 3D QSAR steric coefficients. (a) One latent variable, (b) two latent variables, and (c) six latent variables from the models summarized in Table 10.21.

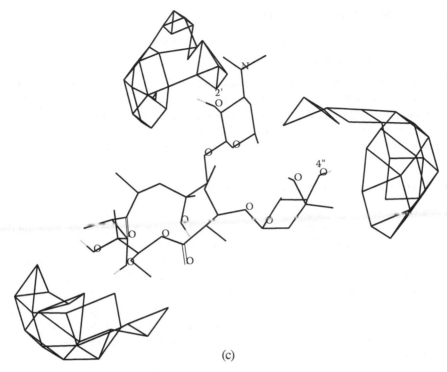

(c)

FIGURE 10.6 *(Continued)*

explain the data. They also emphasize that although the different 2D
and 3D approaches start with different theoretical viewpoints, because
they examine the same structure–activity data, they are intrinsically
related.

TABLE 10.22
Standard Deviations of Prediction Errors of Erythromycin QSAR Models

Model	Analogs Included in the Model	Type of Model	Number of Predictions	Standard Deviation of Predictions
9.5	Alkyl esters, $n = 28$	Linear in log P	44	0.91
10.26	Alkyl esters plus five more polar esters, $n = 33$	Parabolic in log P	39	0.45
10.40	Alkyl esters, $n = 28$	1 latent variable CoMFA 3D QSAR steric	44	0.51
10.41	Alkyl esters, $n = 28$	2 latent variable CoMFA 3D QSAR steric	44	0.53
10.45	Alkyl esters, $n = 28$	6 latent variable CoMFA 3D QSAR steric	44	0.56

V. D1 DOPAMINE AGONISTS

(Collaborators: R. Schoenleber, J. Kebabian, R. MacKenzie,
P. Pavlik Hutchins, J. Wu, C. T. Lin)

The goal of this project was to discover an agonist of the D1 dopamine receptor that did not bind to the α_2 adrenergic receptor nor to the D2 dopaminergic receptor. Such a compound would be used to probe the biological effects of occupying the D1 receptor, and it might lead to a treatment for Parkinsonism.[40] Although the selective D1 agonist SKF-38393, Structure 10.8, was known, it is not a full agonist.[41]

A. MOLECULAR MODELING OF α_2 ADRENERGIC LIGANDS

Prior to our interest in dopamine agonists, we had modeled conformationally constrained analogs of dopamine that bind strongly to the α_2 adrenergic receptor.[42] Structures 10.9 and 10.10 are two of the most potent analogs. To model the conformations of these molecules, we generated candidate conformations with distance geometry[43] and minimized the resulting structures with MMP2.[44] Although both molecules are somewhat conformationally flexible, their low-energy conformations superimpose well, as shown by the distances indicated in Figure 10.7 and the side view shown in Figure 10.8. The rotation of the phenolic hydrogen atoms was selected to conform to that for the homologous D2 dopaminergic receptor discussed in Chapter 3, Figure 3.6.

This superposition provided the necessary 3D information to propose the bioactive conformation of 32 additional diverse agonists[45] and derive a CoMFA 3D QSAR model. We generated this from fields calculated at 1Å spacing, rather than the default 2Å. The CoMFA model for α_2 binding affinity has three components, an R^2 of 0.81, s of 0.54, q^2 of 0.62, and a cross-validated s of 0.73. Only steric fields are statistically significant. Although the s value is not as low as one would like, we considered that it had enough signal that we could use it to forecast the affinity of proposed molecules. In particular, we used the model to rule out molecules with high forecast affinity for the α_2 receptor.

B. MOLECULAR MODELING OF D2 DOPAMINE AGONISTS

As discussed in Chapter 3, a literature ligand-based pharmacophore model[46,47] guided our modeling of D2 dopamine agonists. Structures 10.11 and 10.12 illustrate

(10.8)   (10.9)   (10.10)

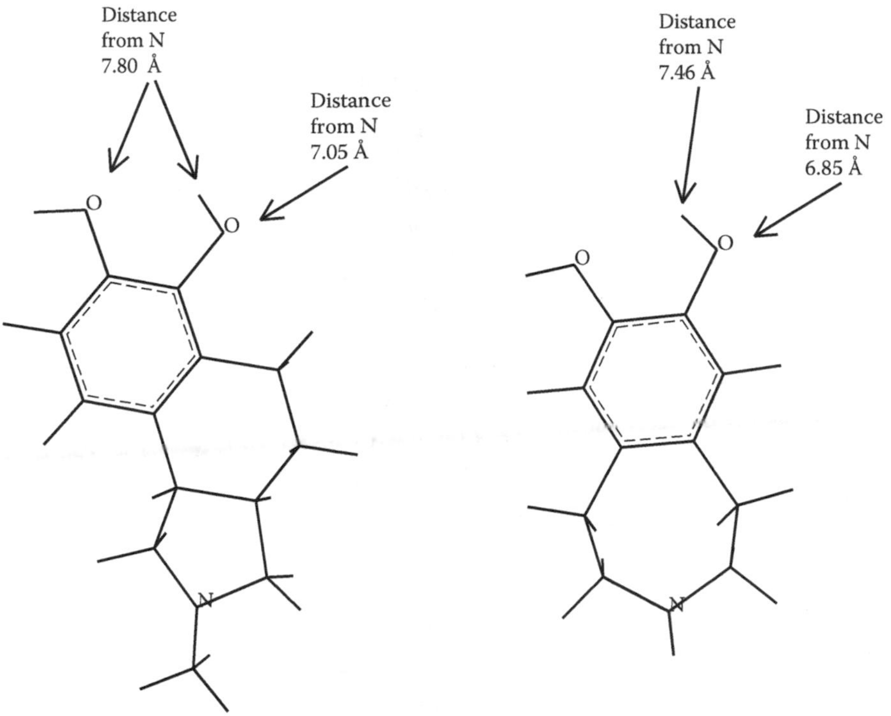

FIGURE 10.7 The distances between key atoms of two conformationally constrained α_2 adrenergic agonists in their global minimum energy conformation.

some of the structure variations that were modeled. Conformations were generated with the same methods that were used for the α_2 agonists. A five-component steric CoMFA 3D QSAR model of the affinity data on a set of 26 catechols and phenols, generated from a lattice of 1Å spacing, had an R^2 of 0.930, an s of 0.401, a q^2 of 0.60, and a cross-validated s of 0.96.[48] We used this model to rule out molecules with high forecast affinity for the D2 dopamine receptor.

C. EXPLORATION OF THE BIOACTIVE CONFORMATION OF PHENYL-SUBSTITUTED CATECHOL AMINES AS D1 DOPAMINE AGONISTS

This work started with the observation that the pendant phenyl of Structure 10.8, SKF-38393, increases affinity for the D1 dopamine receptor by 100-fold over that of Structure 10.10. Molecular modeling of Structure 10.8 with the methods used for the α_2 and D2 agonists suggested that there are two families of low-energy conformations: one with the phenyl group axial and one with it equatorial (Figure 10.9). Note that because of the symmetry of the catechol benzazepine and because the stereochemistry of the diphenyl carbon is known, the conformers superimpose as shown. In the conformers shown in Figure 10.9, the chiral hydrogen atom is above the plane of the molecule.

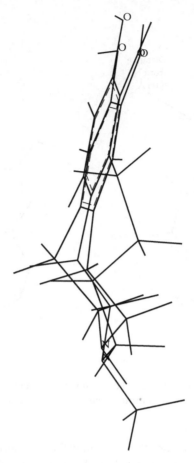

FIGURE 10.8 A side view of the superposition of the two α_2 adrenergic agonists in their global minimum energy conformation.

We decided to probe the location of the binding pocket for the phenyl group by synthesizing mimics of the axial and equatorial conformation using pairs of molecules with and without the added phenyl group. The molecules with the added phenyl groups that we synthesized are shown in Figures 10.10 and 10.11.[49] Although none

(10.11)   **(10.12)**

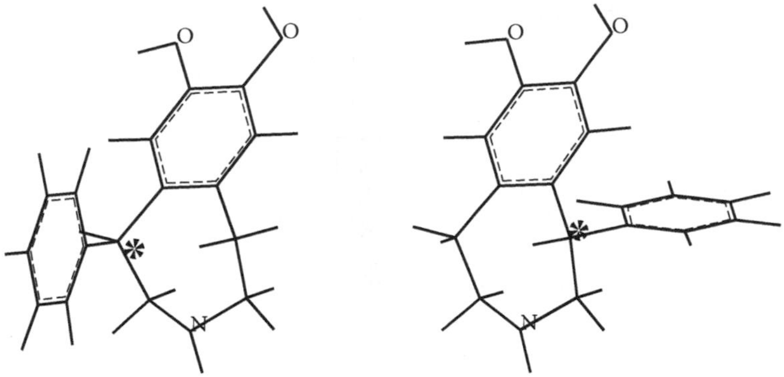

phenyl group axial   phenyl group equatorial

FIGURE 10.9 The two low-energy conformers of SKF-38393.

of the synthesized molecules have high affinity, only the molecules for which the added phenyl is in a position that mimics that in equatorial conformer of SKF-38393 show increased affinity compared to the corresponding analog without the added phenyl. Thus, synthesis of conformationally informative molecules identified the bioactive conformer of SKF-38393 and the relative location of the binding pocket for the phenyl group.

Mimics of the axial conformation of SKF-38393

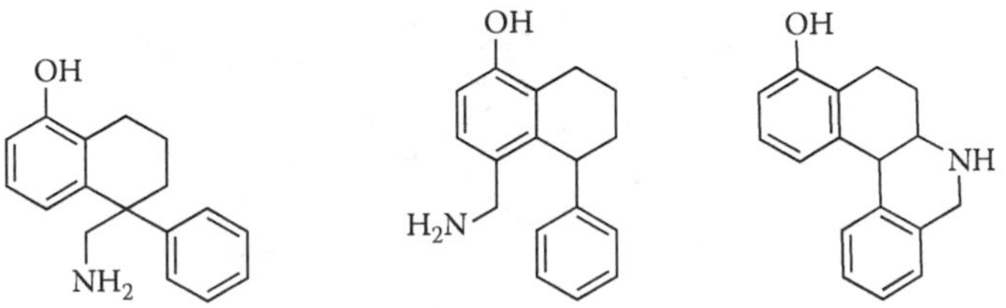

(10.13)   (10.14)   (10.15)

Mimics of the equatorial conformation of SKF-38393

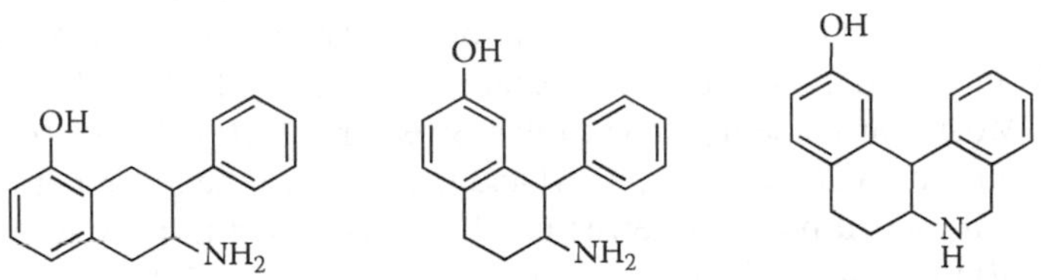

(10.16)   (10.17)   (10.18)

FIGURE 10.10 The structures of molecules synthesized to probe the location of the phenyl-binding pocket on the D1 dopamine receptor.

Mimics of the axial conformation

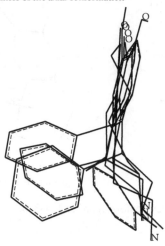

Mimics of the equatorial conformation

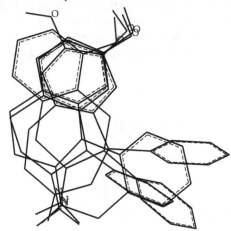

FIGURE 10.11   The 3D superpositions of the mimics of the axial and equatorial conformations of SKF-38393. For clarity the hydrogen atoms are not shown.

As part of our search for novel D1 agonists, project biochemists had measured the D1 affinity of some of the catechol amines from the Abbott collection. Structure 10.19 attracted our attention because its dopaminergic activity had not been previously reported. Molecular modeling showed that Structure 10.20 (Figure 10.12) provides the optimal position of an added phenyl and that it should be given high priority for synthesis. The added phenyl in Structure 10.20 increased affinity by 22-fold over Structure 10.19 to have affinity equal to that of SKF-38393.[49] Additionally, it is a full D1 agonist. Figure 10.12 suggests that these molecules have a very different shape from those that bind to the D2 or the α_2 receptors (Figures 3.5 and 3.6 from Chapter 3, and Figure 10.7).

10.19, R = H
10.20, R = Phenyl

10.21, R = Phenyl

(10.19–10.20) **(10.21)**

Subsequent 3D database searching identified the molecule in which the five-membered ring of Structure 10.19 is opened to form the 1-aminomethyl analog. Adding a phenyl to this molecule also increased affinity. Medicinal chemistry exploration led to the isochroman series,[50,51] for example Structure 10.21, and the observation that the "phenyl-binding pocket" will accommodate groups as large as adamantyl.

D. CoMFA 3D QSAR Predictions of D1 Potency of Novel Catechol Amine Scaffolds

The project team now turned to identifying other structural scaffolds that would correctly position the critical basic nitrogen, phenolic oxygen, and phenyl to produce selective binding to the D1 dopamine receptor. We derived a CoMFA 3D QSAR model for D1 binding, first generating separate models for agonists and antagonists, but after noticing that the resulting contours were similar, combined the two sets

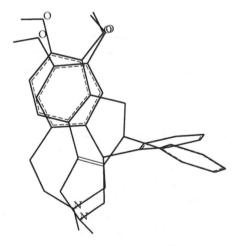

FIGURE 10.12 The superposition of SKF-38393 (Structure 10.8) and Structure 10.20. (Redrawn from Martin, Y. C.; Lin, C. T.; Wu, J. in *3D QSAR in Drug Design. Theory Methods and Applications*; Kubinyi, H., ed. ESCOM: Leiden, 1993, pp. 643–60, Figure 4. With permission.) For clarity the hydrogen atoms are not shown.

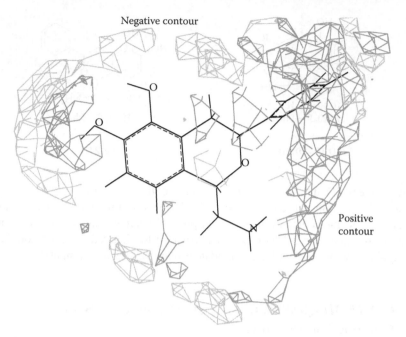

FIGURE 10.13 The CoMFA steric contours for D1 dopamine receptor binding. The darker contours represent regions in space in which it is favorable to place groups and the lighter contours, regions that are unfavorable. (Redrawn from Martin, Y. C.; Lin, C. T.; Wu, J. in *3D QSAR in Drug Design. Theory Methods and Applications*; Kubinyi, H., ed. ESCOM: Leiden, 1993, pp. 643–60, Figure 8. With permission.)

to slightly improve the statistics.[40] This model had an R^2 of 0.96, s of 0.23, q^2 of 0.55, and cross-validated s of 0.79. Only steric contributions were significant. The contours from the CoMFA 3D QSAR (Figure 10.13) highlight the enhanced affinity of molecules that occupy the "phenyl binding pocket." The negative contours below the amino group emphasize that substituting the hydrogens of the amino group with *n*-propyl groups, which increases affinity for the D2 receptor, decreases affinity for the D1 receptor.

The project chemists proposed 201 scaffolds that they deemed synthesizable. For each we generated all stereoisomers and conformers, retained those within 5 kcal/mole of the global minimum, and used the CoMFA models to predict the affinity of each 3D structure for the D1, D2, and α_2 receptors. Of the 146 molecules that the models forecast to have lower affinity for the D1 receptor than dopamine, nine were synthesized and none have higher affinity than dopamine. In contrast, of the 55 molecules that the models forecast to have higher affinity than dopamine, ten were synthesized and six have higher affinity. These six molecules selectively bind to the D1 receptor compared to binding to the D2 and α_2 receptors. Hence, modeling predictions increased the fraction of potent molecules from 0 to 60% and also discouraged synthesis of less potent molecules to 6% of those proposed, compared to 18% of the molecules predicted to have high affinity. An outgrowth of this effort is Structure 10.22, which is active in animal models of Parkinsonism as its diacetyl prodrug, ABT-431.[52–54]

(10.22)

More details on the various CoMFA 3D QSAR studies are reported elsewhere.[40] In particular, we expanded the models to 215 molecules and also explored the CoMFA 3D QSAR of individual series. The statistics and the contours from these various models are very similar to those reported here.

E. LESSONS LEARNED FROM MOLECULAR MODELING OF DOPAMINERGIC AND ADRENERGIC LIGANDS

- The synthesis of conformationally defined ligands led to the establishment of the 3D requirements for affinity for the D1 dopamine receptor.
- The 3D properties of conformationally defined ligands led to the establishment of the differences in the shapes of the binding sites in the D1 and D2 dopaminergic receptors and the α_2 adrenergic receptors, which in turn led to the development of ligands selective for the D1 receptor.
- Predictions from 3D QSAR CoMFA models enriched the fraction of high-affinity ligands in the synthesized molecules.

VI. USE OF LIGAND–PROTEIN STRUCTURES FOR CoMFA 3D QSAR

A. CoMFA ANALYSES TO COMPARE WITH LITERATURE STRUCTURE-BASED QSARs (1997)

(Collaborator: Soaring Bear)

This work was not published. Six publications by others presented QSAR results in which the molecular descriptors were calculated from the protein–ligand complex.[55–60] These calculations are summarized in Table 10.23.

We analyzed the same data using CoMFA. For these studies the structures were minimized with the Sybyl force field and charges on the atoms were calculated with AM1. The conformations were chosen from the experimental protein–ligand complex or modeled from the closest structure. CoMFA steric and electrostatic fields were calculated with the defaults except that 1 Å spacing was used. In five of the six cases, CoMFA produced better cross-validation statistics than were found with the structure-based methods (Figure 10.14). The authors of one of the original papers reported similar findings later.[67]

TABLE 10.23
Literature QSARs Based on Descriptors Calculated from the Protein–Ligand Complex

Enzyme	Number of Observations	Method Used to Calculate the Molecular Descriptors
Carbonic anhydrase[55]	20	Total interaction energy calculated from Amber[61] energy minimization of the ligand, plus the first and second layers of the enzyme. New parameters for zinc interactions.
HIV protease[59]	33	Total interaction energy calculated from MM2X force-field[59] energy, minimization of the ligand in the fixed active site.
Thrombin[56]	35	Total interaction energy calculated from CHARMm force-field[62] energy minimization of the ligand in the fixed active site. Electrostatics were scaled down or neglected during the energy minimization. The interaction energy was calculated with a distance-dependent dielectric function.
Thrombin[60]	7	Coulombic and van der Waals energy calculated using the linear response (LIE) approach[63] using the OPLS force field.[64] (Note: somewhat better statistics are obtained if the solvent-accessible surface area is included; this term is not part of the standard LIE method.)
Neuraminidase[58]	24	Molecular dynamics/energy minimization using CVFF[65] Energies computed, including solvation effects as the change in total electrostatic energy of complexation, using a continuum-electrostatics approach.
Herpes simplex I thymidine kinase[57]	18	The structures were relaxed with Amber[61] and interaction energies calculated with DELPHI.[66]

B. CoMFA ANALYSIS OF UROKINASE INHIBITORS (2007)

Recently, colleagues showed that Molecular Mechanics with Poisson–Boltzmann Surface Area (MMPBSA) calculations fit the affinity of 18 inhibitors of urokinase with an R of 0.90, R^2 of 0.81.[68] To evaluate the performance of CoMFA 3D QSAR on the same data set, we used the experimental structure of the bound ligands as extracted from the complexes with the protein superimposed. No conformational minimization was used. Table 10.24 lists the structure–activity relationships of the ligands, and Table 10.25, the PLS statistics for the CoMFA steric fit. A CoMFA model using two latent variables fits the data substantially better than the more elaborate MMPBSA calculations, and a one-component model fits the data as well as the more elaborate calculations. The three-component model is probably overfit.

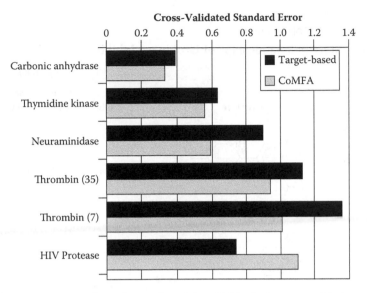

FIGURE 10.14 A comparison of target based fits of biological potency with those fits from CoMFA.

TABLE 10.24
Structure–Activity Relationships of Urokinase Inhibitors

Number	Structure	Log $(1/K_i)$			
		Obs.	Calc. 1 LV	Calc. 2 LV	Calc. 3 LV
(10.23)		7.34	6.77	7.05	7.00
(10.24)		7.15	7.35	7.44	7.16

(Continued)

TABLE 10.24 (CONTINUED)
Structure–Activity Relationships of Urokinase Inhibitors

		Log ($1/K_i$)			
Number	Structure	Obs.	Calc. 1 LV	Calc. 2 LV	Calc. 3 LV
(10.25)		6.49	6.22	6.41	6.64
(10.26)		6.20	5.28	5.78	6.17
(10.27)		6.00	5.45	5.98	6.09
(10.28)		5.75	5.72	5.49	5.66
(10.29)		5.67	6.07	5.66	5.61

(Continued)

TABLE 10.24 (CONTINUED)
Structure–Activity Relationships of Urokinase Inhibitors

Number	Structure	Obs.	Calc. 1 LV	Calc. 2 LV	Calc. 3 LV
(10.30)		5.64	5.33	5.30	5.71
(10.31)		5.50	6.00	5.60	5.60
(10.32)		5.21	4.99	5.46	5.17
(10.33)		5.11	5.58	5.28	5.22
(10.34)		5.00	5.16	4.95	5.15

(Continued)

Log $(1/K_i)$

TABLE 10.24 (CONTINUED)
Structure–Activity Relationships of Urokinase Inhibitors

Number	Structure	Obs.	Calc. 1 LV	Calc. 2 LV	Calc. 3 LV
(10.35)		4.88	5.21	4.86	4.81
(10.36)		4.74	5.28	4.90	4.75
(10.37)		4.53	4.92	5.24	4.78
(10.38)		4.31	4.20	4.14	4.09
(10.39)		4.30	4.07	4.04	4.01
(10.40)		4.00	4.26	4.24	4.18

The "Log $(1/K_i)$" header spans the four data columns (Obs., Calc. 1 LV, Calc. 2 LV, Calc. 3 LV).

TABLE 10.25
Statistics for the CoMFA 3D QSAR Models of the
Urokinase Inhibitors

Latent Variables	1	2	3
R^2	0.80	0.92	0.97
s	0.44	0.30	0.18
q^2	0.59	0.71	0.75
s_{pred}	0.62	0.54	0.53

C.   LESSONS LEARNED FROM STRUCTURE-BASED 3D QSARS

Although these fits do not necessarily indicate that CoMFA 3D QSAR predictions of the affinity of new molecules will be as accurate as those with the more elaborate methods, the ease and speed of CoMFA calculations recommends them.[69–79] The related methods COMBINE[80,81] and AFMoC[82,83] share many of the advantages of CoMFA.

VII.   LESSONS LEARNED

Computational chemists provide alternative viewpoints to the experimentally based team members. Although predictions of potency of as-yet-untested analogs is often important, the computational chemists often suggest molecules that will help distinguish between alternative hypotheses for the property responsible for biological activity: Examples were shown of erythromycin analogs designed to probe the octanonal–water log *P* associated with Gram-positive and Gram-negative antibacterial activity and also of conformationally defined analogs that probe the location of an accessory binding site on the D1 dopaminergic receptor. Preliminary evidence suggests that 3D QSAR may more accurately predict binding affinity than do more elaborate calculations on the ligand–protein complex.

REFERENCES

1. Suda, H.; Takeuchi, T.; Nagatsu, T.; Matsuzaki, M.; Matsumoto, I.; Umezawa, H. *Chem. Pharm. Bull.* **1969**, *17*, 2377–80.
2. Umezawa, H. *Enzyme Inhibitors of Microbial Origin*. University of Tokyo Press: Tokyo, 1972.
3. Umezawa, H.; Takeuchi, T.; Miyano, K.; Koshigoe, T.; Hamano, H. *J. Antibiot.* **1973**, *26*, 189.
4. Hidaka, H.; Asano, T.; Takemoto, N. *Mol. Pharmacol.* **1973**, *9*, 172–7.
5. LogKow/KowWin, Syracuse Research Corporation. Syracuse. 2007. http://www.syrres.com/esc/est_kowdemo.htm (accessed Nov. 1, 2007).
6. Dove, S. *Arch. Pharm.* **2004**, *337*, 645–53.
7. Goldberg, L. I.; Sonneville, P. F.; McNay, J. L. *J. Pharmacol. Exp. Ther.* **1968**, *163*, 188–97.
8. Jones, P. H.; Biel, J. H.; Ours, C. W.; Klundt, I. J.; Lenga, R. L. Amino Acid Amides of Dopamine as Orally Effective Renal Vasodilators 1973 Abstracts of Papers of the American Chemical Society 165th National Meeting MEDI009.

9. Minard, F. N.; Cain, J. C.; Grant, D. S.; Ours, C. W.; Jones, P. H. *Biochem. Med.* **1974**, *11*, 318–26.

10. Kyncl, J.; Martin, Y. C.; Riley, K.; Ours, C. W., Compositions and Methods of Increasing Renal Blood Flow with Upsilon-glutamyl Amide of Dopamine. U.S. Patent 3,947,590, 1975.

11. Kyncl, J.; Riley, K.; Martin, Y. C.; Ours, C. W., Upsilon-glutamyl Amide of Dopamine. U.S. Patent 3,903,147, 1975

12. Jones, P. H.; Ours, C. W.; Biel, J. H.; Minard, F. N.; Kyncl, J.; Martin, Y. C. Gamma-glutamyl Amides of Dopamine as Selective Renal Vasodilators 1976 Abstracts of Papers of the American Chemical Society 172nd Meeting MEDI:17.

13. CLOGP, version 4.3. Biobyte Corporation. Claremont, CA. 2007. http://biobyte.com/bb/prod/40manual.pdf (accessed Nov. 12, 2007).

14. ADME Boxes, Pharma Algorithms. Toronto. http://pharma-algorithms.com/adme_boxes.htm.

15. CoMFA, Tripos, L.P. St. Louis MO. http://tripos.com/data/SYBYL/QSARwCoMFA_072505.pdf (accessed Oct. 16, 2008).

16. CONCORD, Tripos. St. Louis MO. http://tripos.com/data/SYBYL/Concord_072505.pdf.

17. Kimura, E. T.; Dodge, P. W.; Young, P. R.; Johnson, R. P. *Arch. Intl. Pharmacodyn. Ther.* **1971**, *190*, 124–34.

18. Ehrenson, S.; Brownlee, R. T. C.; Taft, R. W. *Prog. Phys. Org. Chem.* **1973**, *11*, 1–80.

19. Hansch, C.; Leo, A.; Hoekman, D. *Exploring QSAR: Hydrophobic, Electronic, and Steric Constants.* American Chemical Society: Washington, DC, 1995.

20. Topliss, J. G.; Edwards, R. P. *J. Med. Chem.* **1979**, *22*, 1238–44.

21. Bird, A. E.; Nayler, J. H. C. In *Drug Design*; Ariens, E. J., Ed. Academic Press: New York, 1971, Vol. 2, pp. 277–318.

22. Lien, E. J.; Hansch, C.; Anderson, S. M. *J. Med. Chem.* **1968**, *11*, 430–41.

23. Pfister, P.; Jenni, S.; Poehlsgaard, J.; Thomas, A.; Douthwaite, S.; Ban, N.; Böttger, E. C. *J. Mol. Biol.* **2004**, *342*, 1569–81.

24. Mao, J. C.-H. In *Drug Action and Drug Resistance in Bacteria. 1. Macrolide Antibiotics and Lincomycin*; Mitsuhashi, S., Ed. University of Tokyo Press: Tokyo, 1971, pp. 153–75.

25. Tardrew, P. L.; Kenney, D. *Appl. Microbiol.* **1969**, *18*, 159–65.

26. Iwasa, J.; Fujita, T.; Hansch, C. *J. Med. Chem.* **1965**, *8*, 150–3.

27. Omura, S.; Katagiri, M.; Umezawa, I.; Komiyama, K.; Maekawa, T.; Sekikawa, K.; Matsumae, A.; Hata, T. *J. Antibiot.* **1968**, *21*, 532–8.

28. Toju, H.; Omura, S. In *Drug Action and Drug Resistance in Bacteria. 1. Macrolide Antibiotics and Lincomycin*; Mitsuhashi, S., Ed. University of Tokyo Press: Tokyo, 1971, pp. 267–91.

29. Martin, Y. C.; Lynn, K. R. *J. Med. Chem.* **1971**, *14*, 1162–6.

30. Monro, R. E.; Fernandez-Munoz, R.; Celma, M. L.; Vazquez, D. In *Drug Action and Drug Resistance in Bacteria. 1. Macrolide Antibiotics and Lincomycin*; Mitsuhashi, S., Ed. University of Tokyo Press: Tokyo, 1971, pp. 305–36.

31. Magerlein, B. J.; Birkenmeyer, R. D.; Kagan, F. *J. Med. Chem.* **1967**, *10*, 355–59.

32. Fujita, T.; Iwasa, J.; Hansch, C. *J. Am. Chem. Soc.* **1964**, *86*, 5175–80.

33. Martin, Y. C. In *Physical Chemical Properties of Drugs*; Yalkowsky, S. H., Sinkula, A. A., Valvani, S. C., Eds. Marcel Dekker: New York, 1980, pp. 49–110.

34. Mason, D. J.; Dietz, A.; DeBoer, C. *Antimicrob. Agents Chemother.* **1963**, 1962, 554.

35. Leo, A.; Jow, P. Y. C.; Silipo, C.; Hansch, C. *J. Med. Chem.* **1975**, *18*, 865–8.

36. Garrett, E. R.; Heman-Ackah, S. M.; Perry, G. L. *J. Pharm. Sci.* **1970**, *59*, 1448–56.

37. Martin, Y. C.; Jones, P. H.; Perun, T.; Grundy, W.; Bell, S.; Bower, R.; Shipkowitz, N. *J. Med. Chem.* **1972**, *15*, 635–8.

38. Iwasaki, H.; Sugawara, Y.; Adachi, T.; Morimoto, S.; Watanabe, Y. *Acta Crystallogr., Sect. C: Cryst. Struct. Commun.* **1993**, *49*, 1227–30.

39. TRIPOS, Inc.: St Louis MO.

40. Martin, Y. C.; Lin, C. T.; Wu, J. In 3D *QSAR in Drug Design. Theory Methods and Applications;* Kubinyi, H., Ed.; ESCOM: Leiden, 1993, pp. 643–60.

41. Setler, P. E.; Sarau, H. M.; Zirkel, C. L.; Saunders, H. L. Eur. *J. Pharmacol.* **1980**, *50*, 419–30.

42. Hancock, A. A.; Kyncl, J. J.; DeBernardis, J. F.; Martin, Y. C. *J. Recept. Res.* **1988**, *8*, 23–46.

43. Crippen, G., Blaney, J., Dixon, S. DGEOM. Distance Geometry; QCPE 590, Quantum Chemistry Program Exchange, Indiana University. Bloomington IN. 1990.

44. Yuri, Y.; Allinger, N. L. *J. Comput. Chem.* **1987**, *8*, 581–603.

45. DeBernardis, J. F.; Kerkman, D. J.; Winn, M.; Bush, E. N.; Arendsen, D. L.; McClellan, W. J.; Kyncl, J. J.; Basha, F. Z. *J. Med. Chem.* **1985**, *28*, 1398–404.

46. McDermed, J. D.; Freeman, H. S.; Ferris, R. M. In *Catecholamines; Basic and Clinical Frontiers;* Usdin, E., Ed. Pergamon: New York, 1979, Vol. 1, pp. 568.

47. Seeman, P.; Watanabe, M.; Grigoriadis, D.; Tedesco, J. L.; George, S. R.; Svensson, U.; Lars, J.; Nilsson, G.; Neumeyer, J. L. *Mol. Pharmacol.* **1985**, *28*, 391–9.

48. Martin, Y. C.; Lin, C. T. In *The Practice of Medicinal Chemistry*; Wermuth, C. G., Ed. Academic Press: London, 1996, pp. 459–83.

49. Martin, Y. C.; Kebabian, J. W.; MacKenzie, R.; Schoenleber, R. In *QSAR: Rational Approaches on the Design of Bioactive Compounds*; Silipo, C., Vittoria, A., Eds. Elsevier: Amsterdam, 1991, pp. 469–82.

50. DeNinno, M. P.; Schoenleber, R.; Asin, K. E.; MacKenzie, R.; Kebabian, J. W. *J. Med. Chem.* **1990**, *33*, 2948–50.

51. Schoenleber, R. W.; Kebabian, J. W.; Martin, Y. C.; DeNinno, M. P.; Perner, R. J.; Stout, D. M.; Hsiao, C.-N. W.; DiDomenico Jr., S.; DeBernardis, J. F.; Basha, F. Z.; Meyer, M. D.; De, B., Dopamine Agonists. U.S. Patent 4,963,568, 1990.

52. Michaelides, M. R.; Hong, Y.; DiDomenico, S., Jr.; Asin, K. E.; Britton, D. R.; Lin, C. W.; Williams, M.; Shiosaki, K. *J. Med. Chem.* **1995**, *38*, 3445–7.

53. Shiosaki, K.; Jenner, P.; Asin, K. E.; Britton, D. R.; Lin, C. W.; Michaelides, M.; Smith, L.; Bianchi, B.; Didomenico, S.; Hodges, L.; Hong, Y. F.; Mahan, L.; Mikusa, J.; Miller, T.; Nikkel, A.; Stashko, M.; Witte, D.; Williams, M. *J. Pharmacol. Exp. Ther.* **1996**, *276*, 150–60.

54. Rascol, O.; Blin, O.; Thalamas, C.; Descombes, S.; Soubrouillard, C.; Azulay, P.; Fabre, N.; Viallet, F.; Lafnitzegger, K.; Wright, S.; Carter, J. H.; Nutt, J. G. *Ann. Neurol.* **1999**, *45*, 736–41.

55. Menziani, M. C.; De, B. P. G.; Gago, F.; Richards, W. G. *J. Med. Chem.* **1989**, *32*, 951–6.

56. Grootenhuis, P. D.; van Galen, P. J. *Acta Crystallogr., Sect. D: Biol. Crystallogr.* **1995**, *51*, 560–6.

57. De Winter, H.; Herdewijn, P. *J. Med. Chem.* **1996**, *39*, 4727–37.

58. Taylor, N. R.; von Itzstein, M. *J. Comput.-Aided Mol. Des.* **1996**, *10*, 233–46.

59. Holloway, M. K. In *3D QSAR in Drug Design. Recent Advances;* Kubinyi, H., Folkers, G., Martin, Y. C., Eds. Kluwer: Dordrecht, 1997, pp. 63–84.

60. Jones-Hertzog, D. K.; Jorgensen, W. L. *J. Med. Chem.* **1997**, *40*, 1539–49.

61. Pearlman, D. A.; Case, D. A.; Caldwell, J. W.; Ross, W. S.; Cheatham, T. E.; Debolt, S.; Ferguson, D.; Seibel, G.; Kollman, P. *Comput. Phys. Commun.* **1995**, *91*, 1–3.

62. Brooks, B. R.; Bruccoleri, R. E.; Olafson, B. D.; States, D. J.; Swaminathan, S.; Karplus, M. *J. Comput. Chem.* **1983**, *4*, 187–217.

63. Åqvist, J.; Medina, C.; Samuelsson, J.-E. *Protein Eng.* **1994**, *7*, 385–91.

64. Jorgensen, W. L.; TiradoRives, J. *J. Am. Chem. Soc.* **1988**, *110*, 1657–66.

65. Dauber-Osguthorpe, P.; Roberts, V. A.; Osguthorpe, D. J.; Wolff, J.; Genest, M.; Hagler, A. T. *Proteins: Struct., Funct., Genet.* **1988**, *4*, 31–47.

66. Honig, B.; Nicholls, A. *Science* **1995**, **268**, 1144–9.

67. Bursi, R.; Grootenhuis, P. D. *J. Comput.-Aided Mol. Des.* **1999**, *13*, 221–32.

68. Brown, S. P.; Muchmore, S. W. *J. Chem. Inf. Model.* **2007**, *47*, 1493–503.

69. Diana, G. D.; Kowalczyk, P.; Treasurywala, A. M.; Oglesby, R. C.; Pevear, D. C.; Dutko, F. J. *J. Med. Chem.* **1992**, *35*, 1002–8.

70. Waller, C. L.; Oprea, T. I.; Giolitti, A.; Marshall, G. R. *J. Med. Chem.* **1993**, *36*, 4152–60.

71. Kroemer, R. T.; Ettmayer, P.; Hecht, P. *J. Med. Chem.* **1995**, *38*, 4917–28.

72. Dajani, R.; Cleasby, A.; Neu, M.; Wonacott, A. J.; Jhoti, H.; Hood, A. M.; Modi, S.; Hersey, A.; Taskinen, J.; Cooke, R. M.; Manchee, G. R.; Coughtrie, M. W. H. *J. Biol. Chem.* **1999**, *274*, 37862–8.

73. Matter, H.; Schwab, W.; Barbier, D.; Billen, G.; Haase, B.; Neises, B.; Schudok, M.; Thorwart, W.; Schreuder, H.; Brachvogel, V.; Lonze, P.; Weithmann, K. U. *J. Med. Chem.* **1999**, *42*, 1908–20.

74. Sippl, W.; Holtje, H. D. *J. Mol. Struct.* **2000**, *503*, 31–50.

75. Pintore, M.; Bernard, P.; Berthon, J.; Chretien, J. R. *Eur. J. Med. Chem.* **2001**, *36*, 21–30.

76. Bhattacharjee, A., K.; Geyer, J., A.; Woodard, C., L.; Kathcart, A., K.; Nichols, D., A.; Prigge, S., T.; Li, Z.; Mott, B., T.; Waters, N., C. *J. Med. Chem.* **2004**, *47*, 5418–26.

77. Sciabola, S.; Carosati, E.; Baroni, M.; Mannhold, R. *J. Med. Chem.* **2005**, *48*, 3756–67.

78. Wei, H. Y.; Tsai, K. C.; Lin, T. H. *J. Chem. Inf. Model.* **2005**, *45*, 1343–51.

79. Verma, R. P.; Hansch, C. *Bioorg. Med. Chem.* **2007**, *15*, 2223–68.

80. Ortiz, A. R.; Pisabarro, M. T.; Gago, F.; Wade, R. C. *J. Med. Chem.* **1995**, *38*, 2681–91.

81. Ortiz, A. R.; Pisabarro, M. T.; Gago, F.; Wade, R. C. In *QSAR and Molecular Modelling: Concepts, Computational Tools and Biological Applications*, Proceedings of the 10th European Symposium on Structure-Activity Relationships: QSAR and Molecular Modelling, Barcelona, September 4–9, 1994; Sanz, F., Giraldo, J., Manaut, F., Eds.; J. R. Prous: Barcelona, 1995, pp. 439–43.

82. Gohlke, H.; Klebe, G. *J. Med. Chem.* **2002**, *45*, 4153–70.

83. Silber, K.; Heidler, P.; Kurz, T.; Klebe, G. *J. Med. Chem.* **2005**, *48*, 3547–63.

11 Methods to Approach Other Structure–Activity Problems

This chapter highlights techniques to investigate problems related to QSAR analysis that cannot be addressed with a least-squares analysis:

How does one design a set of molecules whose biological properties will be modeled with QSAR?

How can one include inactive molecules or develop models for biological responses that are not continuous, but rather, categorical, such as active/inactive or mutagenic/not mutagenic?

How can one analyze data sets that contain diverse molecular structures?

How can one display the relationships between molecules in property space?

How can one decide if new molecules are similar enough to those in the training set that the predictions will be accurate?

The techniques described in this chapter represent extensions of the concepts of principal components analysis and partial least squares that were introduced in Chapter 7. The basis of these techniques is the concept that molecules are located at a point in the multidimensional space defined by the molecular descriptors used. The techniques differ in a number of ways that make each especially suited to some specific use.

I. MEASURING THE SIMILARITY OR DISTANCES BETWEEN MOLECULES

Molecular similarity (or its converse, distance) is a fundamental concept of these methods. It is calculated from the properties of each molecule and an algorithm that specifies how it will use the differences in these properties between pairs of molecules. There are many choices in both the properties to be used and the algorithm for the calculation.[1] This discussion will highlight only those most commonly used.

In QSAR, every molecule is associated with the values of a number of molecular properties. These values determine the location of the molecule in a space that has the dimension of the number of properties. Thus, if our molecules are described by five different properties, each molecule is a point in a five-dimensional space, and its location along each axis is determined by the value of the property corresponding to that axis. If it is described by a vector of 256 properties, it is a point in a 256-dimensional space.

Although usually only a few dimensions are required to describe the relative positions of the molecules in property space, the maximum number of possible dimensions in property space is equal to the lesser of the number of properties or the number of molecules considered. For example, the maximum dimensions of a data set of 1000 molecules characterized by 250 descriptors is 250. However, the maximum dimensions of a data set of 100 molecules characterized by 250 descriptors is 100.

A.   DISTANCE CALCULATIONS BASED ON MOLECULES DESCRIBED BY CONTINUOUS PROPERTIES

In a typical QSAR, each molecular property is a continuous variable; that is, it can assume any value within the range of that property. However, each property has its own range; for example, in a set of substituents for an aromatic ring, the range of π is from -4.36 to 2.66 (or 7.02 units), but that of MR is 0.92 to 65.51 (or 64.59 units), and that of σ_m is -0.84 to 0.93 (or 1.77 units). This variation in ranges means that a 1.0 unit difference in the MR of two molecules is a minor difference, but the same difference in σ_m is dramatic. As a result of the variation in the property ranges, one must decide how much relative influence the various properties will have on the distance or similarity calculation. Will the distances be calculated in terms of the units of the original measurements, differences in π, MR, and σ_m, for example, or will they be calculated so that each measurement contributes equally to the distance criterion?

Frequently, workers calculate a standardized or normalized property value by subtracting the mean and dividing by the standard deviation of that property. An example of the influence of weighting variables is seen by a consideration of the relative magnitudes of π and MR for the -H, -benzyl, and -ethylcarbamate ($-NHC(=O)OC_2H_5$) substituents on an aromatic ring (Table 11.1). Note that the ethylcarbamate is closer to -H in terms of π, but closer to benzyl in terms of MR. From the mean and standard deviation of π and MR in a large series of substituents, we calculated the standardized values shown in the table. Ethylcarbamate is still closer to benzyl in terms of MR, but now the difference is smaller. Note that the standardized values will depend on the population that was used to calculate the mean and standard deviation.

Many other possible weighting schemes may be used. For example, one might weight the value of a property by subtracting the median and dividing by the range of the observations: Such weighting is less sensitive to the influence of individual points than normalization. Different weighting schemes may or may not lead to different conclusions; that depends on the data set involved.

A different weighting scheme is needed if the values of some of the molecular properties are correlated. For example, if molecules were described by log P values calculated by three different programs, but by only one steric and electrostatic property, the correlated properties (the various calculated log P values) would increase

TABLE 11.1

Example of the Effect of Standardization on Distance Measures

Substituent Formula	-H	-NHC(=O)OC₂H₅	-CH₂C₆H₅
π	0.00	0.17	2.01
MR	1.03	21.18	30.01
Standardized π	−0.66	−0.50	1.15
Standardized MR	−1.10	0.25	0.85
Euclidean distance: raw values	20.15	0.00	9.02
Hamming distance: raw values	20.32	0.00	10.67
Euclidean distance: standardized values	1.36	0.00	1.76
Hamming distance: standardized values	1.51	0.00	2.25

the hydrophobicity contribution to the distance compared to the contribution of the steric and electrostatic properties. This is seen in Table 7.6, which lists a set of molecules for which log P has been calculated with two different programs and log D with another. For data sets that contain correlated variables, a principal components analysis reveals how many independent properties or dimensions are present. Table 7.8 shows that for the data in Table 7.6 there are at most three significant principal components. The resulting principal component scores of the molecules, (Table 7.10) contain the appropriate weighting of the original properties.

Once the weighting scheme is selected, one must decide how to calculate the distance between two molecules A and B. The Euclidean distance (Equation 11.1) is commonly used; it is the square root of the sum of squares of the differences between the values of the variables, p_i.

$$E_{A,B} = \sqrt{\sum (p_{i,A} - p_{i,B})^2} \tag{11.1}$$

Alternatively, the Hamming or city block method, Equation 11.2, simply sums up the absolute values of the differences:

$$H_{A,B} = \sum |p_{i,A} - p_{i,B}| \tag{11.2}$$

Table 11.1 shows that, when compared using Euclidean or Hamming distances, ethylcarbamate is closer to—more similar to—benzyl if the raw values are used, but closer to hydrogen if standardized values are used. With standardized values the distinction between the distance to benzyl and hydrogen becomes much smaller.

Distances can be calculated in a number of additional ways. For example, the distance measure might be the cube root of the sum of cubes of the differences, or the fourth root of the sum of the fourth power of the differences, or the mean of the distances in each dimension. The higher the order of the exponential applied to the difference, the more the calculation weights large differences in one dimension only. Again, the choice of method may or may not influence the conclusions that one would draw from a study.

B. DISTANCE CALCULATIONS BASED ON MOLECULES DESCRIBED BY A VECTOR OF 1'S AND 0'S

Frequently, one describes molecules with a long vector of 1's and 0's to indicate the presence or absence of a particular feature. Hence, each molecule occupies a point in that high-dimensional space. Similarity of molecules in such a space is typically represented by the Tanimoto coefficient. It is calculated by Equation 11.3 from the number of bits set in common, c, and the number of bits set in Molecule A and Molecule B, a and b, respectively:

$$T_{A,B} = \frac{c}{a+b-c} \qquad (11.3)$$

Note that Tanimoto distance would be $1 - T_{A,B}$. Many other similarity measures are also available for this type of data.[1,2]

II. DISPLAYING LOCATIONS OF MOLECULES IN MULTIDIMENSIONAL SPACE

Examining the locations of the molecules in property space is an important part of developing a QSAR model: Are the molecules in the data set evenly dispersed? Do some molecules form clumps of similar molecules, or are some molecules distant from the others? Clumps, if large enough, might form the basis for a local QSAR analysis; a more dispersed set of molecules might provide a QSAR that has a better ability to forecast the potency of diverse molecules; and molecules that are far removed from the remainder might be outliers from a QSAR or they might unduly influence an apparent relationship. Although in principle one could examine the locations of molecules in each pair of properties, for even 10 properties this would result in 45 plots. Furthermore, such pairwise plots do not reveal unsuspected correlations between several descriptors.

The choice of molecular descriptors affects the locations of molecules in the calculated space. Using physical properties as molecular descriptors supports QSAR analyses, but using substructure descriptors, as described in Chapter 4, provides a basis for understanding the structural diversity of a data set. Because these descriptors are complementary, both are useful.

From Chapter 7 we learned that, although it is possible to describe molecules with many properties, in fact, these properties may not be independent. For example, the nine properties of the molecules in Table 7.6 reduce to two because of (1) the correlations among the various calculations of log P and correlations between these log P descriptors and calculated solubility, molecular weight, and number of rotatable bonds; and (2) the correlation between polar surface area and calculated log brain/blood ratio. To take advantage of such relationships within a data set, statisticians and computer scientists have devised a number of methods to project the multidimensional locations of objects—molecules in our case—into fewer dimensions. Chapter 7 described the use of two of these methods, principal components[3] and partial least squares,[3–5] for small data sets. Table 11.2 summarizes the characteristics of these and some of the other common methods.[6–11] All support visualization by providing methods that combine the properties into fewer descriptors that capture most of the original information.

The contribution of molecular properties to such composite descriptors provides an opportunity to efficiently investigate how the various properties are related to each other and to external properties. This information might lead to omitting certain properties from the analysis because they do not vary, they are highly correlated with other properties, or they are not correlated with the target property. To investigate such relationships, principal components and factor analysis are the appropriate tools.

Principal components and factor analysis both extract linear combinations of properties using the correlation matrix as the basis, but differ in the exact calculation.[10] The principal components of a data matrix are extracted to maximize the explained variance in the molecular properties. Hence, the first components contain contributions (loadings) of many of the properties. However, this can confuse the interpretation of the relationships between the components and the properties. Factor analysis makes the relationships easier to interpret because it rotates the principal components in the multidimensional space to maximize the contribution of some properties while minimizing the contributions of others.[12–14] In practice, a principal components analysis is often run first to establish the number of significant or interesting components, then factor analysis is run to produce the same number of factors. By analogy to principal components, the factors are described by the loadings (contributions) of the properties that contribute to that factor, and factor scores are calculated for each observation.

For example, we calculated the principal components and factors of the physical properties of a set of molecules for which human bioavailability has been reported in the literature.[15] Figures 11.1 and 11.2 show the principal components and factor score plots. The points are colored by the value of their CLOGP[16]-calculated octanol–water log P. Although both plots show the contribution of log P to the scores, log P influences the values of the first two principal components but only the value of Factor 1. The plots emphasize that it is easier to interpret the meaning of the factors than of the principal components.

The distribution of the molecules in property space is also of interest. The unshaded version of Figures 11.1 and 11.2 would show this. However, because principal components and factor analysis maximize the variance in properties extracted, they are sensitive to the relative density of points (molecules) at a certain region in space—increasing the number of molecules at a certain region puts more weight on this region in the component or factor. As a result, the distances between molecules in principal component or factor space do not scale with their Euclidean distances in property space.

To attempt to maintain the distance between objects—molecules in our case—one would use a multidimensional scaling or Kohonen neural networks program

TABLE 11.2

A Summary of Methods to Reduce the Dimensionality of a Data Set

Method	What It Optimizes	Special Uses
Principal components analysis, PCA[3]	Linear combinations of molecular properties that maximize the explained variance in these properties.	1. Estimate the number of dimensions in a table of molecular properties. 2. Use principal component loadings as descriptors for QSAR. 3. Use the loadings for variable selection if potency was included in the PCA. 4. Examine the location of molecules in multidimensional space.
Factor analysis[6,10,12]	Linear combinations of molecular properties that maximize the explained variance in these properties, rotated to maximize or minimize the loadings of individual properties.	1. Use factor scores as descriptors for QSAR. 2. Use the loadings for variable selection if potency was included in the factor analysis. 3. Examine the location of molecules in multidimensional space.
Partial least squares[3,4,5]	Linear combinations of molecular properties that maximize the explained variance in the dependent property.	1. Build QSAR models using many molecular properties that are intercorrelated. 2. Examine the location of molecules in multidimensional space.
Multidimensional scaling[8,9,11]	Linear and nonlinear combinations of molecular properties that preserve the (usually Euclidean) distances between pairs or multiples of molecules emphasizing recreating long distances.	Examine the location of molecules in multidimensional space.
Kohonen neural network[7,8]	Linear and nonlinear combinations of molecular properties that preserve the linear and curved relationships between the molecules in multidimensional space emphasizing recreating short distances.	Examine the location of molecules in multidimensional space.

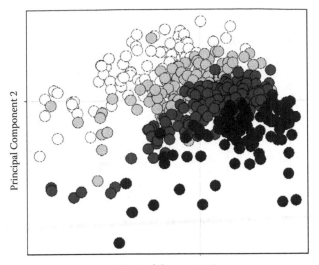

Principal Component 1

FIGURE 11.1   Plots of the principal component scores calculated from drugs for which bio-availability in the human has been reported. The calculations used the following molecular properties: predominant charge at pH 6, log P calculated with CLOGP, log P calculated with KowWin, count of H bond donors, count of nitrogen plus oxygen atoms, molecular weight, and polar surface area. The open circles correspond to molecules for which the log P calculated with CLOGP is less than or equal to 0.0; light gray molecules for which log P is greater than zero but less than or equal to 2.0; dark gray, log P is greater than 2.0 but less than or equal to 4.0; black, log P is greater than 4.0.

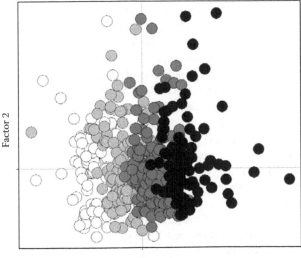

Factor 1

FIGURE 11.2   Plots of the factor scores for the same data as in Figure 11.1. The factors were generated with the varimax algorithm. The points are shaded as in Figure 11.1.

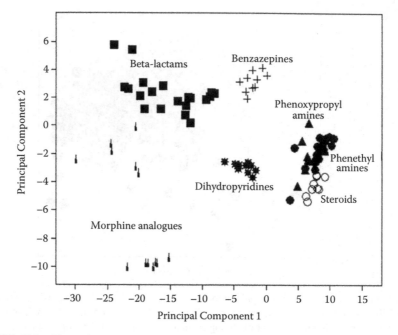

FIGURE 11.3 Plots of the principal component scores calculated from drugs for which bio-availability in the human has been reported. The calculations are based on the Daylight finger-prints of length 2048. Only the compounds in the seven largest clusters are shown (Table 11.3).

to reduce the dimensionality of the data.[11,17] These example calculations use the same molecules used for Figures 11.1 and 11.2, but describe the molecules with the Daylight fingerprints,[18] which encode the 2D structures of the molecules. Although all the molecules were used for the calculations, Figures 11.3 and 11.4 show only the molecules in the seven largest clusters of similar molecules. The figures illustrate the subtle differences between the projections of molecules using principal components and multidimensional scaling calculated with the R option in PipeLine Pilot.[19] Note that, in the multidimensional scaling plot, the clusters of the molecules are tighter and more separated from the other clusters. For example, the individual beta lactam antibiotics and the individual benzodiazepines are closer to one another in MDS space but more dispersed in principal component space. Although neither method totally separates the phenoxypropyl and phenethyl amines, multidimensional scaling provides hints of separation.

Figures 11.3 and 11.4 also illustrate how one might select molecules for a QSAR analysis. Individual clumps of molecules could be used to derive a local QSAR for related molecules, whereas if one wanted a global QSAR that might cover any molecule, then molecules should be selected from all areas of the multidimensional scaling space. The figure also indicates that certain regions of this space are more heavily populated—for a global model, it would be important to not sample this region more heavily than more sparse regions.

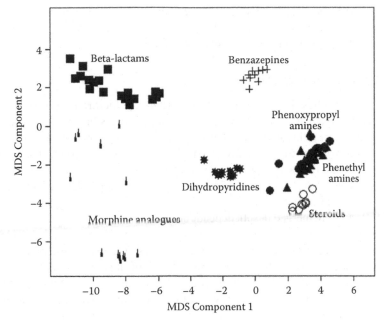

FIGURE 11.4 Plots of the multidimensional scaling scores calculated for drugs for which bio-availability in the human has been reported. The calculations are based on the Daylight finger-prints of length 2048. Only the compounds in the seven largest clusters are shown (Table 11.3).

III.   GROUPING SIMILAR MOLECULES TOGETHER

Cluster analysis uses the calculated pairwise similarities of molecules to partition them into groups or clusters that have similar properties but are different from the members of other clusters. The results of a cluster analysis depend on three choices made by the user: the properties of the molecules used in the calculations; the equation used to calculate distances between molecules; and the specific algorithm used to form the clusters. In contrast to regression, partial least-squares or principal components analysis; cluster analysis does not describe a single algorithm, but rather a family of algorithms.[20–23]

Cluster analysis can answer two types of questions: "Which of these observations are similar to each other?" and "How many different groups does this data represent?" For example, cluster analysis on a large set of aromatic substituents established which subsets of the total were similar.[24] If one wished to synthesize a series of molecules with the maximum uncorrelated variation in the properties, then one would choose one member from each cluster.

One can also hypothesize that because the molecules within a cluster have similar properties, if one is biologically active, the others in the cluster are also expected to be active. On this basis, if one wanted to select a subset of molecules for biological testing, one would chose one or a few from each cluster. The problems with this use are that (1) not all properties may be associated with biological activity, (2) including correlated variables may weight the clusters inappropriately, and (3) clustering is imprecise. Nevertheless, molecules that are similar to an active are more likely to themselves be active than are randomly chosen molecules.[25]

A. HIERARCHICAL CLUSTERING

As the name implies, hierarchical methods produce a hierarchy of how individual molecules group at every cluster level from 1 (all molecules in one cluster) to n (each molecule in an individual cluster). Clustering may be agglomerative, starting from one cluster per observation, and progressively coalescing the two most similar clusters; or it may be divisive, starting with all of the observations as one large cluster and progressively forming more clusters. In successive steps, the decisions made in previous steps are not changed.

How clusters are merged or split depends on how one calculates the distance between multimember clusters: The nearest neighbor or single linkage method considers this distance to be that between the two closest members of the two clusters; the centroid method considers it to be that between the centers of the clusters; and the complete linkage method considers it to be that between the two most distant members of the two clusters. No clustering method is perfect; for example, the single linkage method can produce clusters in which some molecules are very different from each other, whereas the complete linkage method can put two similar molecules into different clusters.

Hierarchical clustering methods output both the hierarchy of the order in which clusters are grouped or partitioned as well as the associated similarities for forming each cluster. Abrupt changes in the distance needed to join or split clusters helps one decide how many homogeneous clusters are present in the data. These methods also output a one-dimensional ordering of the molecules according to how they clustered.

A simple example of a cluster dendogram is shown in Figure 11.5. It shows the molecules from Table 7.6 clustered by the first two principal components of the properties

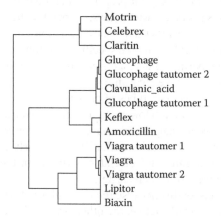

FIGURE 11.5 The dendogram produced by clustering the first two principal components of the data in Table 7.9. The clustering was performed with the complete linkage method and Euclidean distances.

in Table 7.9. Compare this figure with Figure 7.3, which shows the locations of the molecules in the space of the first two principal components. In the dendogram, the distance to the branching point is proportional to the distance between new clusters formed. Thus, the properties of the molecules in the cluster of Motrin, Celebrex, and Claritin are very different from the other molecules but not especially different from each other. Because it has the longest line (the largest distance) between it and the molecules with which it clusters, the dendogram shows that Biaxin has the most unique complement of physical properties. Although the clustering reproduces in a general way the relationships in Figure 7.3, it does not accurately show relationships such as the following: Lipitor is approximately as close to Celebrex as it is to Biaxin, and it is farther away from Keflex than from Celebrex.

B. PARTITIONAL CLUSTERING

Partitional clustering distributes the members of a data set into a single set of clusters, not a hierarchy of clusters. These methods generally run faster than hierarchical methods. The downside is that they provide no information about the relationships between clusters.

The k-means method[20,26,27] is an example of a method that makes a predetermined number of clusters n. The algorithm starts by selecting the user-specified n seed molecules at random, assigns the remaining molecules to the most similar seed molecule, and calculates the centroids (the geometric means) of these initial clusters. Any molecule that is closer to the centroid of a different cluster than to the initial seed is relocated to that cluster. The centroid calculation-relocation process is repeated until no changes in cluster membership occur or until some predefined limit of the number of iterations is reached. Because the initial seed molecules are selected randomly, the k-means method does not produce a unique solution.

The Taylor method makes clusters with a predetermined between-molecule similarity.[25,28,29] It was preferred by a group of Abbott medicinal chemists who examined the ability of various algorithms to cluster active molecules identified by high-throughput screening. The algorithm starts by identifying for every molecule all other molecules within the user-specified similarity limit. The first iteration identifies the first cluster to contain the molecule that has the largest number of similar molecules and all of the molecules within its similarity limit. The next iteration removes this cluster of molecules from the list, and repeats the process until no more clusters are possible. Although all the molecules within a cluster are similar to the lead molecule at the chosen level, a problem with this algorithm is that certain molecules may be outside the similarity limit of the lead molecule but are similar to other molecules in the cluster. Such molecules may be rescued, but this is at the expense of the strict similarity criterion.

The Taylor algorithm was used to cluster the human bioavailability data set of 513 molecules. The molecules were described with the Daylight 2D fingerprints described in Chapter 4. A Tanimoto similarity of 0.65 was used to identify the molecules similar to each input structure. The result is 276 clusters, 79 of these having more than one molecule and 197 singletons. Examples of the seven largest clusters are shown in Table 11.3. The human bioavailability of each molecule is also shown. These seven clusters include 19% of the molecules.

TABLE 11.3

Example Structures from Each of the Seven Largest Clusters in the Human Bioavailability Data Set

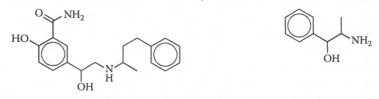

Cluster 1, Beta-lactams, *n* = 20

Cefoperazone, 5%F   Cefoxitin, 78%F

Cluster 2, Phenethyl amines, *n* = 19

Labetalol, 18%F   Phenylpropanolamine, 80%F

Cluster 3, Phenoxypropyl amines, *n* = 14

Penbutolol, 90%F   Propranolol, 26%F

Cluster 4, Morphine analogs, *n* = 13

Hydromorphone, 52%F   Vincamine, 20%F

(*continued*)

TABLE 11.3 (CONTINUED)
Example Structures from Each of the Seven Largest Clusters in the Human Bioavailability Data Set

Cluster 5, Dihydropyridines, $n = 11$

Isradipine, 19%F

Nifedipine, 50%F

Cluster 6, Steroids, $n = 11$

Prednisone, 4%F

Canrenoic acid, 4%F

Cluster 7, Benzodiazepines, $n = 10$

Clorazepate, 95%F

Desmethyldiazepam, 99%F

Two Similar Molecules That Do Not Cluster Together

Benzylpenicillin, Cluster 1, 22%F

Penicillin V, Cluster 8, 60%F

Note: The fraction bioavailable is indicated next to the name of the structure.

Table 11.3 also shows, at the bottom of the table, two similar molecules, one from Cluster 1 and one from Cluster 8. This highlights a potential problem with clustering. Because every clustering method can split related molecules and many make clusters of unrelated molecules, if one uses clustering to produce subsets that will be used for QSAR, it is important to use these clusters as only a preliminary step to selecting the subsets.

IV. ANALYZING PROPERTIES THAT DISTINGUISH CLASSES OF MOLECULES

A. GENERAL INTRODUCTION

Sometimes the objective of a structure–activity analysis is not to predict the relative potency of active molecules, but rather to predict whether a particular molecule will be active or inactive or whether it will be an agonist, an antagonist, or inactive. One might use such an analysis to analyze high-throughput screening results or to build a model from a set of structurally diverse molecules. Although simple variants of regression analysis and partial least squares may be used for this purpose, several other algorithms are available.[27,30–34]

The classification methods discussed in this section will be demonstrated with a data set of 1623 molecules that had been tested in the late 1950s in a search for a novel inhibitor of mitochondrial monoamine oxidase.[35] We hypothesized that inhibiting this enzyme would increase the levels of norepinephrine and serotonin, and so alleviate the symptoms of depression. In this program, compounds were screened at a rate of 24 or 48 per week with the potency reported as inactive, slightly active, active, and potent—Classes 0–3, respectively, in the examples that follow. Each week's set of molecules contained those newly synthesized for the program, analogs of molecules that the medicinal chemist identified as similar to those that were active in previous testing, and diverse compounds manually chosen to broadly sample the amines available in the compound collection. No computer software was used for the selection of compounds for testing.

As a result of the methods used to add compounds for testing, the data set contains several series of related molecules embedded within the larger set of structurally diverse molecules. Table 11.4 shows an example from each of the eight largest clusters.

For the analyses reported in this section, the molecules were prepared by removing counterions from all structures except quaternary salts. In addition, compounds that are complexes of transition metals were removed from the data set. All comparisons were performed using the FPFC_6 circular fingerprints of SciTegic,[36] ignoring stereochemistry. Recall that these fingerprints describe each atom by its atom type and the bond type and atom type of all atoms within three bonds from it.

Table 11.5 illustrates a confusion matrix, also called a contingency table or a truth table. This is a common method to condense the results of a classification model. The column at the right of the table shows the recovery rate, the percentage of the molecules that belong a particular class that are fitted to that class. It shows that the model discovers 1523 (99%) of the 1535 inactive molecules and 55 (62%) of the 88 active molecules. The row at the bottom of the table shows the percentage of the

TABLE 11.4
One Example from Each of the Largest Clusters of Active Inhibitors in the MAO Data Set

Cluster	Number of Members	Example Structure
1	43	
2	20	
3	15	
4	17	
5	11	
6	6	
7	5	
8	5	

molecules fitted to a particular class that actually belong to that class: Thus, 98% of the molecules that were fitted to be inactive are inactive, but only 82% of the molecules that were fitted to be active were active. Although the overall identification of the correct class is 97%, this is dominated by the fact that 99% of the inactive molecules are recovered, while only 62% of the active molecules are recovered. Of course, in an ideal model the cross-validation values would match the fitted ones, and such a model would recover 100% of the active molecules while not incorrectly predicting any of the inactive molecules to be active.

TABLE 11.5

An Example Confusion Matrix Summary of a Classification Analysis

	Fitted Inactive	Fitted Active	Total	% Recovered
Actual inactive	1523	12	1535	99
Actual active	33	55	88	62
Total	1556	67	1623	
Correct	1523	55	1578	
% Correct	98 (specificity)	82 (sensitivity)	97	

Viewed another way, the table shows that if one tested only the molecules that are predicted to be active, then only 67, or 5% of the whole set, would need to be tested, but 62% of the actives would be found. This represents an enrichment of $(55/67)/(88/1623) = 15$ fold. Other analyses of this data set will be discussed below.

If a calculation involves only two classes, a positive or active and a negative or inactive class, for example, then sometimes the model results are presented as sensitivity and specificity. Sensitivity is the probability that a positive prediction is measured to be positive—that a molecule predicted to be active is indeed active, a true positive. The complement is the specificity, which is the probability that a negative prediction is measured to be negative—that a molecule predicted to be inactive is indeed inactive, a true negative. A perfect model would thus have specificity and sensitivity equal to 1.0. In the model summarized in Table 11.5, the sensitivity is 82% and the specificity is 98%. Note that although the specificity is 98%, the model classifies 33 active molecules into the inactive class.

Often, the numbers of observations in the separate classes are not equal. This is reflected in the prior probability of a class, which is often calculated as the ratio of the number of members of a class to the total number of observations. Based on chance alone, one would predict a new observation to belong to the class with the highest prior probability. In the example, the prior probability of the inactive class is 0.94 (1535/1623) and of the active class is 0.06. On the other hand, predictions from a model are termed *posterior probability*. The various algorithms may use the prior probabilities as part of the calculation of posterior probabilities.

The prior probabilities need not be proportional to the size of the classes used for the calculation, but rather they can be generated from another assessment of the relative frequencies of the two classes in the total population. For example, sometimes the number of inactive molecules is so large that a modeling algorithm cannot detect the features of the active molecules. The solution to this problem is to use a sample of the inactive molecules but all of the active molecules to build the model. In this case, one might use the prior probabilities from the total population.

In addition to considering prior probabilities, some algorithms also consider the relative cost of misclassification of objects from the two classes. For example, increasing the cost of misclassification of the active class would lead to a model in which more inactive molecules would be classified as active but more active molecules would be identified, an increase in recovery. On the other hand, increasing the cost of misclassification of inactive molecules would lead to a model in which more active molecules would be classified as inactive but those predictive to be active would be more likely to be active—an increase in sensitivity.

B. EXAMPLES OF CLASSIFICATION METHODS

1. Linear Discriminant Analysis (LDA)

Discriminant analysis is the classical tool to identify the features that differentiate classes of molecules.[31–34] It is closely related to regression analysis. Whereas regression analysis calculates the contributions of the predictor properties to best fit the value of an observed dependent property, discriminant analysis calculates the contributions of the predictor properties to best distinguish the classes of molecules.

Recall from Chapter 7 that in regression analysis, the contributions of the predictor properties are calculated to maximize the fraction of the total variance in the dependent property that is explained. In discriminant analysis, the contributions of the predictor properties are calculated to maximize the ratio of the between-class variance to the corresponding within-class variance. In fact, one can calculate the simplest form of two-class discriminant analysis by using 1.0 and −1.0 as the dependent property in a regression or partial least-squares program and using a forecast potency of 0.0 as the cutoff point between classes.

The aim of a linear discriminant analysis investigation is to discover which combination of predictor properties leads to the best discrimination between the classes of molecules. By analogy with regression analysis, this discovery can involve a stepwise or all-possible-models approach.

Linear discriminant analysis is not limited to two classes; indeed, the classic introduction of this method separated three classes of iris on the basis of the morphology of their leaves and flowers.[31] An implied assumption of discriminant analysis is that the members of each class can be separated from those of the other classes by a straight line in multidimensional property space. In other words, it assumes that the molecules of a certain class are not surrounded in property space by molecules of a single other class.

Recall that regression analysis finds the line that best fits potency in the multidimensional space of the properties of the molecules. In a similar way, discriminant analysis finds the plane that best separates the classes in the multidimensional space of the properties.

There can be subtleties in how the calculations indicate the class to which an observation is predicted to belong. In the simplest case, the classes to be distinguished are approximately equal in size and the cost of misclassification from any class is equal. In such an instance, the computer program computes the mean value of the discriminant function for each class, and the point half-way between these means is used as the cutoff point between classes. Each observation is assigned a predicted class by comparing the value of its calculated function with the cutoff point.

Using prior probabilities in discriminant analysis has the effect of moving the boundary plane that separates the classes closer to the class with the higher prior probability. Because there are more observations in this class, it is assumed that the properties of the class are better defined and that for a new observation to be predicted to belong to this class, its properties should be within the domain of the members of the class. Hence, using prior probabilities changes the predicted class of a molecule to the lower-probability class if its properties are different from those of the higher-probability class. The posterior probability is then the calculated probability that the observation belongs to the class in question, considering both the prior probability and the values of the predictor variables.

In practice, for a discriminant analysis calculation one uses a standard statistical program such as R.[37,38] If one is using physical properties for the analysis, before the calculation is run one should eliminate any variables that have the same mean for the different classes. The molecular properties may be used as such or one may use the principal component scores as the descriptors. In addition to the matrix of predictor variables for each observation, one must usually specify the prior probability of each class.

A special case of discriminant analysis, quadratic, is used when the covariances ("spread") of the individual classes are not equal. This is the case if one class is more restricted in space than the others. Discriminant analysis programs will warn the user if quadratic discriminant analysis is necessary for a particular data set. If so, the computer program performs the usual analysis but does not pool the covariance matrices.

Discriminant analysis is especially suited to modeling high-throughput screening data, because the potency of the active compounds is usually not precisely known. One would consider a discriminant function valuable if it can identify the members of the lower-frequency class. In the example shown in Table 11.5, although the active molecules represent 5.4% of the data set, the fitted discriminant function enriches this to 82% of the molecules assigned to the active class. As noted above, this represents an enrichment of 15-fold and recovers 62% of the active molecules.

Table 11.6 summarizes the results of discriminant analysis for different two-class groupings of the monoamine oxidase data set. In the table, the analysis is performed for three divisions of the data set: (1) Set 1, active molecules are those 88 molecules with activity rating 3 and all others are classified as inactive, (2) Set 2, active molecules are those 174 molecules with activity rating 2 or 3 and all others are classified as inactive, and (3) Set 3, active molecules are those 286 molecules with activity rating 1, 2, or 3, and those with activity rating 0 are classified as inactive. The table shows that as less potent molecules are added to the class of actives (moving from Set 1 to Set 3), the discriminant function is able to recover a lower percentage of the actives; the percentage drops from 63% to 28%.

As with any QSAR method, it is not correct to evaluate a discriminant function with the same data that was used to form it. Although a training and test set should have been used, the leave-one-out cross-validation (LOO CV) statistics in Table 11.6 suggest that the functions appear to be robust. For example, even in the function based on Set 3, the function enriched the actives from 21% to 53%.

In the case of an active synthetic or iterative screening program, it may be useful to use the original (small) set of active molecules to form a discriminant function, verify the function by its ability to predict the potency of new analogs as they are tested, and then revise the function to include the newly tested molecules. This process would be repeated until adding more molecules no longer improves the discriminant function.

TABLE 11.6
A Comparison of Discriminant Analysis, Support Vector Machine, K-Nearest Neighbors, and Recursive Partitioning of the Monoamine Oxidase Data Using SciTegic FPFC_6 Fingerprints as Descriptors

	Predicted Active	Found Active	% Actives in Predicted Actives (100X Sensitivity)	% Actives Recovered
Set 1: Active Is Defined as Highly Potent: 88 Active and 1535 Inactive, 5.7% Active				
Discriminant analysis, fit	67	55	82	62
Support vector machine, fit	59	57	97	65
K-nearest neighbor, k = 3	85	62	72	70
Recursive partitioning, with boosting, fit	193	83	43	94
Recursive partitioning, top of first tree (2 splits)	64	58	91	66
Discriminant analysis, LOO CV	69	49	71	56
Support vector machine, 10-fold CV	50	47	94	53
Recursive partitioning, LOO CV	149	69	46	78
Recursive partitioning, 10-fold CV	111	65	59	74
Set 2: Active Is Defined as Moderately or Highly Potent: 174 Active and 1449 Inactive, 12.1% Active				
Discriminant analysis, fit	95	76	80	44
Support vector machine, fit	76	76	100	44
K-nearest neighbor, k = 3	140	89	64	51
Recursive partitioning, with boosting, fit	511	164	32	94
Recursive partitioning, top of first tree (3 nodes)	77	69	90	40
Discriminant analysis, LOO CV	97	67	69	39
Support vector machine, 10-fold CV	61	58	95	33
Recursive partitioning, LOO CV	380	113	30	65
Recursive partitioning, 10-fold CV	342	109	32	63
Set 3: Active Is Defined as Any Activity: 286 Active and 1337 Inactive, 21.4% Active				
Discriminant analysis, fit	95	81	85	28
Support vector machine, fit	94	94	100	33
K-nearest neighbor, k = 3	233	118	51	41
Recursive partitioning, with boosting, fit	1302	283	22	99
Recursive partitioning, top of first tree (6 nodes)	117	99	85	35
Discriminant analysis, LOO CV	122	65	53	23
Support vector machine, 10-fold CV	66	61	92	21
Recursive partitioning, LOO CV	963	232	24	81
Recursive partitioning, 10-fold CV	1356	269	20	94

In summary, discriminant analysis is a useful extension to regression analysis because it allows one to include inactive molecules in the analysis and to analyze nonquantitative data of other types. It is closely related to regression analysis in that in discriminant analysis, the function minimized is the sum of the squared deviations of observations from their respective class means, while the sum of squares from the opposite class is maximized; in regression analysis, the function minimized is the sum of the squared deviations of observations from a line.

2. SIMCA

Recalling that molecules are located in a multidimensional space, one can imagine that molecules that belong to different classes occupy different regions in this space. Indeed, if they did not occupy different regions, then discriminant analysis could not find a plane to separate the classes. However, it is also possible that the principal components of the various classes are different; for example, a particular property might be constant within the active series but vary within the inactives. A SIMCA analysis calculates the principal components of each class separately and uses the scores of these principal components to calculate the predicted class membership.[3,39] It also calculates the distance of the molecule to the principal component region of the class and rejects predictions that are not within the class boundaries.

The advantage of SIMCA over discriminant analysis is that it forms a separate model of each class of molecules. The result is that if different descriptors characterize the various classes, there will be a better separation. A second advantage is that SIMCA automatically recognizes if a new molecule is an outlier in property space.

3. Support Vector Machine Classifiers (SVMs)

Support vector machines present a new type of machine learning algorithm.[27,40–44] They may be used to fit either continuous data, as in regression analysis or PLS, or they may be used as classifiers, that is, trained to distinguish classes of molecules.

Support vector machine classifiers are a direct analog of discriminant analysis classifiers in that they search for a plane that separates the classes of interest. However, in SVMs this decision plane is calculated to maximize the distance (the margin) between the plane and the nearest molecules in each class. The margin is maximized until it bumps into the some observations, the support vectors. Thus, a SVM model can be visualized as a curved thick slab that separates the classes. This visualization contrasts with a linear discriminant, which describes a plane that separates classes.[40]

Sometimes, any plane in the dimensions of the original data cannot separate the classes. For example, this would occur if the active molecules form a clump in multidimensional space and are surrounded by inactive molecules. In such cases, one can generate nonlinear SVM classification functions by using a kernel function that distorts the separating plane so that it is not flat in the original dimensions but appears flat for the calculations. Besides linear classifiers, typical choices for nonlinear classifiers include quadratic or other polynomial functions or radial basis function machines, but many others have been designed. If the classes are not linearly separable, then the SVM plane is calculated to maximize the margin while minimizing prediction error.

Developing an SVM model requires that one decide the form and parameters of the kernel and the trade-off between misclassifying an observation and maximizing the margin between classes. Usually, one develops multiple models with different parameters, and the one with the best prediction accuracy is used for further predictions.

The superiority of SVM models over discriminant analysis models is seen in Table 11.6. For every set of data the SVM recovered the same number of or more active molecules with fewer false positives. Neither the fitted discriminant function nor the SVM models recover all of the active molecules. Although they cannot be directly compared, the cross-validation statistics suggest that both methods are satisfactory.

In summary, SVM classifiers appear to be useful improvements over discriminant analysis. In particular, Table 11.6 suggests that, although both methods recover a similar fraction of actives, the SVM identified fewer false positives. Two disadvantages are that several runs must be made to identify the options that yield the best model and that the calculations are much slower than discriminant analysis.

4. K-Nearest Neighbor Prediction (KNN)

In contrast to LDA, SIMCA, and SVMs, the K-nearest neighbor method does not assume that all molecules of a class occupy continuous regions in multidimensional space. Rather, KNN predicts activity using only the activities of the most similar molecules in the data set. It assumes that molecules that have similar molecular properties are likely to have similar biological activity and that one can predict the biological activity of an untested molecule by considering the biological activity of molecules that are similar to it.[27,45] As a result, it can make predictions even if molecules of a given class can occupy different islands in multidimensional space.

Usually more than one neighbor, the k in KNN, is used for the prediction: The value of k is empirically chosen. It typically varies from three to nine. Optionally, one may elect to include only molecules within a certain distance or similarity threshold and/or also to weight closer molecules more heavily for the predictions. KNN predictions depend not only on the predictors used for the calculation but also how the distances are calculated, and if and how the predictors are weighted.

Unlike LDA and SVMs, which derive functions that weigh the importance of each predictor, KNN assumes that all predictors are equally important. Clearly, if unimportant predictors are used, then the predictions will not be as reliable as if only relevant ones had been used. Including correlated variables will also bias the distance calculations.

The predictions using KNN for the example data set are shown in Table 11.6. These calculations did not involve weighting of the neighbors by distance. The predictions are true predictions in that the input molecule was not used in the calculation. However, often one assesses the stability of a KNN model by cross-validation, leaving out some of the molecules. Note that, for the example, the KNN model for each of the sets retrieves somewhat more actives than either LDA or SVM but that the fraction of actives within the set of predicted actives is smaller than these two methods.

5. Recursive Partitioning or Decision Trees

The previously described methods use all input molecular properties, perhaps weighted, to classify molecules. In contrast, recursive partitioning methods, RP, use properties one at a time to classify molecules.[46–52] Furthermore, as with KNN, for recursive partitioning it is not necessary that all molecules of a class occupy the same region in space—in fact, the algorithms are designed to detect these different regions.

Recursive partitioning algorithms start by finding the molecular property that best splits the molecules into nodes (subsets) that are enriched in one class or the other. After this split, they search each node individually for a property that further separates the classes. Splitting the nodes continues until some stopping criterion is reached. The result is a tree such as shown in Figure 11.6. Each branch represents a decision point, and each node contains the subset of molecules that have been partitioned to that node. Comparing the decision tree in Figure 11.6 with the clusters members in Table 11.4 shows that the recursive partitioning missed 12 of the compounds from Cluster 1, and all of the molecules in Clusters 3, 6, and 7.

The figure shows an important advantage of recursive partitioning; that is, that the important molecular properties are highlighted: Indeed, one use of recursive partitioning is to identify properties to use in other methods. Because a decision tree is built from a series of decisions, the classification can also be reported as the rules that lead to each node of the decision tree.

Different recursive partitioning algorithms differ in the criteria that decide which property will be used for the split. Typical choices are to select the property with the largest (standardized or not) difference in variance between the two classes, to select the property that produces the purest (highest proportion of one class) nodes, or to use an information-theoretic calculation to select the property to use for the split. As part of the calculation, one can also adjust the relative costs of misclassification.

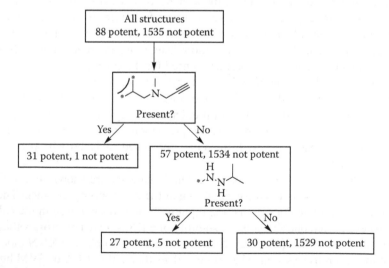

FIGURE 11.6 The first few splits of the recursive partitioning of the MAO data using ECFP_6 descriptors.

The algorithms need to decide when to stop the splitting because, in the extreme case, a tree could be built such that every molecule would be in a node by itself. Typical stopping points are when no split will generate some fractional increase in enrichment in the daughter nodes, when a node has reached some preset minimum size, or when a preset number of levels has been reached.

In addition to removing some of the data to be used as a test set, usually cross-validation is a part of a recursive partitioning calculation. The cross-validation may involve the usual leave-one-out or leave-some-fraction out, but often it also automates using a fraction of the training set to form a validation set that is not used in the training but is used to provide an assessment of overfitting.

A decision tree may not represent the optimum way to partition a data set because the properties are selected to make the splits by considering only the node to be split; they do not have the ability to look ahead to see if another property might lead to a better decision tree. Two different approaches to solve this problem are random forests and boosting. Both generate multiple possible decision trees to classify examples.

In the random forest method,[52] many (usually at least 100) decision trees are built on subsets of the training data. The algorithm chooses the molecules in these subsets by bootstrapping: selecting molecules at random with replacement, that is, with the ability to be chosen more than once for a particular subset. For each node of each tree, a very small random sample of the molecular properties is considered for the split. In the final model, each tree provides a vote as to the class to which a molecule belongs.

In boosting methods, after one tree is built, each subsequent tree is built by reweighting the importance of each observation by its residual from the previous iteration. This gives more weight to poorly fitted molecules. Boosting uses the accuracy of predictions of an independent validation set to decide when to terminate the process. The final prediction is then the weighted average of all the predictions. If boosting is used, it is extremely important to reserve a completely independent test set for model validation.

Table 11.6 highlights one problem with recursive partitioning: although the first few splits may identify features of the minority (active) class, further splits degrade the purity of this class while recovering more of the actives. In the extreme case of Set 3, which contains 21.4% actives, the active nodes in the full recursive partition model would contain 22% actives. For this data set, stopping at six nodes improves fraction of actives to 85%, but at a sacrifice of recovery from 99% to 35%. Hence, it is necessary to use a carefully chosen stopping point with recursive partitioning.

C. OBSERVATIONS FROM ANALYSES OF MONOAMINE OXIDASE INHIBITION DATA SET

Table 11.6 contains only representative results. For many of the methods, improved classification may result from a different choice of calculation options. However, for this data set, SVMs appear to be an improvement over linear discriminant analysis.

KNN, in general, retrieves more actives with a slight decrease in precision, while recursive partitioning can retrieve most of the actives at the expense of a further decrease in specificity.

A particular problem with this particular data set is that it includes several series of active molecules, as shown in Table 11.4. Thus, these series of molecules are easy for the algorithms, just as for a human, to find.

V. LESSONS LEARNED

The similarity of two molecules depends on the descriptors used for the calculation, how the individual descriptors are weighted, and the algorithm used for the similarity calculation.

The visualization of the relationships between molecules is aided by methods that project many properties into a few. Projections calculated by principal components or factor analysis attempt to explain the maximum amount of variance of the molecular properties, whereas those from multidimensional scaling or Kohonen neural networks preserve the multidimensional distances between molecules. Hence, projections of the same data will depend on the calculation algorithm. Plots of the contribution of the molecular properties to the projections reveal which properties of the data set are closely related and which are not.

Cluster analysis groups similar molecules together. The results depend on the molecular descriptors as well as the clustering algorithm. Hierarchical clustering shows how the molecules associate as fewer and fewer clusters are formed. The distance between clusters can help one estimate the number of "natural" clusters. K-means clustering forms a predetermined number of clusters, which supports selecting diverse molecules for biological testing. The Taylor method makes clusters based on a predetermined similarity criterion to provide a measure of the number of "natural" clusters.

Various techniques may be used to develop models that distinguish between categories of data. Discriminant analysis, a direct analog of linear regression analysis, identifies molecular descriptors that form planes that separate the classes, one less plane than the number of classes. Support vector machines are similar but allow the planes to deform to better separate the classes. The k-nearest neighbor methods predict the activity of molecules from the activity of the k most similar molecules; hence, for this method, all molecules of a certain class do not need to be close to each other. Recursive partitioning considers one property at a time that separates molecules into classes and repeats this partitioning, one property at a time, on each previously identified subset of the data. The result is a decision tree that can also be described by a set of rules.

REFERENCES

1. Leach, A. R.; Gillet, V. J. *An Introduction to Chemoinformatics*. Springer: Dordrecht, 2005.
2. Willett, P.; Barnard, J. M.; Downs, G. M. *J. Chem. Inf. Comput. Sci.* **1998**, *38*, 983–96.
3. Eriksson, L.; Johansson, E.; Kettaneh-Wold, N.; Trygg, J.; Wikström, C.; Wold, S. *Multi- and Megavariate Data Analysis. Part I: Basic Principles and Applications*, 2nd ed. Umetrics AB: Umeå, 2007.

4. Bush, B. L.; Nachbar Jr., R. B. *J. Comput.-Aided Mol. Des.* **1993**, *7*, 587–619.
5. Garson, G. D., Partial Least Squares Regression (PLS); 2008. http://faculty.chass.ncsu. edu/garson/PA765/pls.htm (accessed Nov. 6, 2008).
6. Manly, B. F. J. *Multivariate Statistical Methods. A Primer*, 2nd ed. Chapman and Hall: London, 1994.
7. Gasteiger, J.; Holzgrabe, U.; Kostenis, E.; Mohr, K.; Surig, U.; Wagener, M. *Pharmazie* **1995**, *50*, 99–105.
8. Bayada, D. M.; Hamersma, H.; van Geerestein, V. J. *J. Chem. Inf. Comput. Sci.* **1999**, *39*, 1–10.
9. Shi, L. M.; Fan, Y.; Lee, J. K.; Waltham, M.; Andrews, D. T.; Scherf, U.; Paull, K. D.; Weinstein, J. N. *J. Chem. Inf. Comput. Sci.* **2000**, *40*, 367–79.
10. Principal Components and Factor Analysis; 2008. http://www.statsoft.com/textbook/ stathome.html?stfacan.html&1 (accessed April 28, 2008). StatSoft, Inc.
11. Multidimensional Scaling; 2008. http://www.statsoft.com/textbook/stathome.html? stmulsca.html&1 (accessed April 28, 2008). StatSoft, Inc.
12. Darlington, R. B., Factor Analysis; http://www.psych.cornell.edu/Darlington/factor.htm (accessed Jan. 25, 2008).
13. Morrison, D. F. *Multivariate Statistical Methods*. McGraw-Hill: New York, 1990.
14. Anon., Principal Components and Factor Analysis; 2008. http://www.statsoft.com/ textbook/stfacan.html (accessed Jan. 25, 2009). StatSoft, Inc.
15. Martin, Y. C. *J. Med. Chem.* **2005**, *48*, 3164–70.
16. CLOGP, version 4.3. Biobyte Corporation. Claremont, CA. 2007. http://biobyte.com/ bb/prod/40manual.pdf (accessed Nov. 12, 2007).
17. Naud, A., Neural and Statistical Methods for the Visualization of Multidimensional Data; 2001. Uniwersytet Mikokaja Kopernika w Toruniu. http://www.fizyka.umk.pl/ ~naud/01phd-an.pdf (accessed April 24, 2008).
18. Fingerprints–Screening and Similarity; 2008. http://www.daylight.com/dayhtml/doc/ theory/theory.finger.html, accessed: Aug. 3, 2008. Daylight Chemical Information Systems, Inc.
19. Pipeline Pilot, version 5.1.0.100. SciTegic. San Diego, CA. 2005.
20. Cluster Analysis; Wikipedia. http://en.wikipedia.org/wiki/Cluster_analysis (accessed March 26, 2008).
21. SAS/STAT® 9.2 User's Guide. Introduction to Clustering Procedures; http://support. sas.com/documentation/cdl/en/statugclustering/61759/PDF/default/statugclustering.pdf (accessed March 26, 2008). SAS Inc.
22. Everitt, B. S. *Cluster Analysis*. Edward Arnold: London, 1993.
23. Cluster Analysis; 2008. www.statsoft.com/textbook/stathome.html?stcluan.html&1 (accessed March 26, 2008). StatSoft, Inc.
24. Hansch, C.; Unger, S. H.; Forsythe, A. B. *J. Med. Chem.* **1973**, *16*, 1212–22.
25. Martin, Y. C.; Kofron, J. L.; Traphagen, L. M. *J. Med. Chem.* **2002**, *45*, 4350–8.
26. Hartigan, J. A.; Wong, M. A. *Appl. Statistics* **1979**, *28*, 100–8.
27. Wu, W.; Kumar, V.; Quinlan, J. R.; Ghosh, J.; Yand, Q.; Motoda, H.; McLachlan, G. J.; Ng, A.; Liu, B.; Yu, P. S.; Zhou, Z.-H.; Steinbach, M.; Hand, D. J.; Steinberg, D. *Knowledge Information Systems* **2008**, *14*, 1–37.
28. Taylor, R. *J. Chem. Inf. Comput. Sci.* **1995**, *35*, 59–67.
29. Butina, D. *J. Chem. Inf. Comput. Sci.* **1999**, *39*, 747–50.
30. Sherrod, P. H., Linear Discriminant Analysis; http://www.dtreg.com/lda.htm (accessed March 26, 2008). DTREG.
31. Fisher, R. A. *Annals of Eugenics* **1936**, *7*, 179–86.
32. Martin, Y. C.; Holland, J. B.; Jarboe, C. H.; Plotnikoff, N. *J. Med. Chem.* **1974**, *17*, 409–13.
33. Frank, I. E. *Chemom. Intell. Lab. Syst.* **1989**, *5*, 247–56.

34. Discriminant Function Analysis; 2008. http://www.statsoft.com/textbook/stathome. html?stdiscan.html&1 (accessed March 26, 2008). StatSoft, Inc.
35. Taylor, J. D.; Wykes, A. A.; Gladish, Y. C.; Martin, W. B. *Nature* **1960**, *187*, 941–2.
36. SciTegic Chemistry Components 2008. http://accelrys.com/products/datasheets/chemistry-component-collection.pdf (accessed May 4, 2008). Accelrys.
37. The R Project for Statistical Computing, 2008. http://www.r-project.org/ (accessed May 7, 2008).
38. Albers, W., Discriminant Analysis; 2008. http://www.home.math.utwente.nl/~albersw/ Discriminant%20Analysis.ppt (accessed May 6, 2008).
39. Dunn III, W. J.; Wold, S.; Martin, Y. C. *J. Med. Chem.* **1978**, *21*, 922–30.
40. Bennett, K. P.; Campbell, C. *SIGKDD Explorations* **2000**, *2*, 1–13.
41. Byvatov, E.; Fechner, U.; Sadowski, J.; Schneider, G. *J. Chem. Inf. Comput. Sci.* **2003**, *43*, 1882–9.
42. Norinder, U. *Neurocomputing* **2003**, *55*, 337–46.
43. Byvatov, E.; Schneider, G. *J. Chem. Inf. Comput. Sci.* **2004**, *44*, 993–9.
44. Arimoto, R.; Prasad, M. A.; Gifford, E. M. *J. Biomol. Screen.* **2005**, *10*, 197–205.
45. Zheng, W. F.; Tropsha, A. *J. Chem. Inf. Comput. Sci.* **2000**, *40*, 185–94.
46. Hawkins, D. M.; Young, S. S.; Rusinko, A. *Quant. Struct.-Act. Relat.* **1997**, *16*, 296–302.
47. Chen, X.; Rusinko, A.; Young, S. S. *J. Chem. Inf. Comput. Sci.* **1998**, *38*, 1054–62.
48. Young, S. S.; Hawkins, D. M. *SAR QSAR Environ. Res.* **1998**, *8*, 183–93.
49. Dixon, S. L.; Villar, H. O. *J. Comput.-Aided Mol. Des.* **1999**, *13*, 533–45.
50. Jones-Hertzog, D. K.; Mukhopadhyay, P.; Keefer, C. E.; Young, S. S. *J. Pharmacol. Toxicol. Methods* **1999**, *42*, 207–15.
51. Blower, P.; Fligner, M.; Verducci, J.; Bjoraker, J. *J. Chem. Inf. Comput. Sci.* **2002**, *42*, 393–404.
52. Breiman, L. *Machine Learning* **2001**, *45*, 5–32.

Index